D1064138

Women and Gaming

Women and Gaming

The Sims and 21st Century Learning

James Paul Gee and Elisabeth R. Hayes

palgrave
macmillan

WOMEN AND GAMING
Copyright © James Paul Gee and Elisabeth R. Hayes, 2010.

First published in 2010 by PALGRAVE MACMILLAN® in the United States—a division of St. Martin's Press LLC, 175 Fifth Avenue, New York, NY 10010.

Where this book is distributed in the UK, Europe, and the rest of the world, this is by Palgrave Macmillan, a division of Macmillan Publishers Limited, registered in England, company number 785998, of Houndmills, Basingstoke, Hampshire RG21 6XS.

Palgrave Macmillan is the global academic imprint of the above companies and has companies and representatives throughout the world.

Palgrave® and Macmillan® are registered trademarks in the United States, the United Kingdom, Europe and other countries.

ISBN: 978-0-230-62341-5

Library of Congress Cataloging-in-Publication Data

Gee, James Paul.
  Women and gaming: The Sims and 21st century learning / James Paul Gee and Elisabeth R. Hayes.
    p. cm.
  Includes bibliographical references and index.
  ISBN 978-0-230-62341-5
  1. Simulation games in education. 2. Video games—Social aspects. 3. Video games for women. 4. Sims—Computer games. I. Hayes, Elisabeth. II. Title.

LB1029.S53G44 2009
794.8082—dc22   2009038279

A catalogue record of the book is available from the British Library.

Design by Scribe Inc.

First edition: May 2010

10 9 8 7 6 5 4 3 2 1

Printed in the United States of America.

Contents

Acknowledgments

We wish to thank the MacArthur Foundation whose Digital Media and Learning initiative funded the research on which this book is based. We particularly wish to thank Connie Yowell, the director of the initiative. Connie has been a driving force behind a great deal of the research on digital media and learning over the last few years. Her work and vision has been instrumental in starting this new field off on a solid footing. We would also like to thank our former colleagues in what is now the Games, Learning, and Society Program at the University of Wisconsin–Madison: Erica Halverson, Rich Halverson, David Williamson Shaffer, Constance Steinkuehler, and Kurt Squire. So much of our work on games and learning was developed while we were among these outstanding young scholars.

We would also like to thank the women who were part of the Tech-Savvy research team over the three years of the project: Elizabeth King, Barbara Johnson, Jayne Lammers, and Yoonhee Lee. Their commitment to the project and willingness to take on every task, no matter how large or small, were invaluable. We want to recognize in particular Beth King's contribution to Chapter 4. She devoted endless hours to working with Jade and the other girls in the first Tech Savvy Girls group and collected much of the data that formed the basis for that chapter.

Lastly, we wish to express our gratitude to Sam, Jade, Tabby Lou, Alex, and Jesse, whose stories provide the heart and soul of this book. We also thank the other girls and women (and men) who are mentioned in the book, as well as the many others who inspired us and challenged our thinking about women and gaming.

Chapter 1

Introduction

Gaming Goes Beyond Gaming

In this book, we are going to talk about video games. But, perhaps surprisingly, we are going to talk about people doing things with games beyond playing them. We are going to talk about how gaming today has moved beyond just playing the game to involve players with design, production, and participation in learning communities. Furthermore, we are going to focus on girls and women. Though much of the media and academic interest in games has focused on males, we argue that, in the trend to take gaming beyond game play, females are leading the way in some very interesting respects.

Much attention has been paid to violent video games like *Grand Theft Auto* or *Counter-Strike*. However, a great many video games are not violent: games like *SimCity*, *Harvest Moon*, and the beautiful independent game *Flower*. In this book we will concentrate on a nonviolent game that has been controversial, *The Sims*.

The Sims is the best-selling video game in history. In 2008, *The Sims*, in all it versions and expansions, reached a milestone of over one hundred million copies sold since its launch and is still going strong (Ortutay, 2008). *The Sims 3*, which came out in June 2, 2009, sold 1.4 million copies in its first week, the best launch in Electronics Arts' (EA's) history, one of the world's largest game companies (Terdiman, 2009). The game has been published in sixty countries and twenty-two languages (Mountjoy, 2008).

In *The Sims*, players build houses and guide the lives of virtual characters ("Sims") in neighborhoods and communities. *The Sims* has been called many things: a sandbox, a toy, a dollhouse, a story engine, and a virtual world, as well as a game.

The Sims is a game where the majority of players are girls and women, though many males play it as well (Lockwood, 2007). It is odd, perhaps, that when males play a military came like *Call of Duty*, we do not say they are playing with toy soldiers. But when women play *The Sims*, we say they are playing with a dollhouse (Schiesel, 2006). Some hard-core gamers have a hard time respecting *The Sims* and its players (Marken, 2009). Yet this is deeply ironic, since *The Sims* not only is the best-selling game in history but also is made by a game designer, Will Wright, who is revered by nearly all gamers and a good many educators. Wright has designed such games as *SimCity*, *The Sims*, and *Spore*.

For years, Will Wright has exemplified a trend that is now becoming pervasive in gaming. His games involve players alternating between playing the game and designing things. This trend is particularly dramatic in his most recent game, *Spore* (http://www.spore.com). In *Spore*, players start out as single-celled creatures in a pond. They have to survive and flourish in the pond to earn the right to become a land creature. As a land creature, they must engage in a fight for the survival of the fittest with other creatures. If they survive, they eventually become intelligent creatures and begin to form cultures and then civilizations. By the end of the game, players have left Earth on a spaceship and are terraforming planets in space, bringing about new life.

At each stage of *Spore*, players must design and build before they can play, designing their own creatures, buildings, vehicles, and spaceships. The design tools in *Spore* are very user friendly, and players have designed a massive amount of widely creative things. The game is made in such a way that player creations are downloaded, via the Internet, into each player's game, eventually replacing, for the most part, the things designed by Wright and his team of professionals. It is these player-designed creatures, their vehicles and buildings, with which the player must compete. Wright's attitude is that the very best things made by players are often better than the things designed by professionals.

The Sims also involves a good deal of design alongside playing (Hayes, 2008b; Hayes & King, 2009). Players can design their own houses and clothes for their Sims. However, players have also designed or found a good many technical tools for making things that are not available inside the game. Whole communities design things for the game, such as new types of houses, furniture, clothes, and landscapes, and make them available to other players. These communities have become unpaid workers for Will Wright, extending the life of the game while resourcing their own creativity.

The Sims exists in different versions (*The Sims 3* is most current), and earlier versions have many expansion packs that add things to the basic game. Different players play and design for different versions of the game and not all have moved to the newer versions. We will see that *The Sims* has motivated many players to spend less time playing the game (albeit often after hundreds of hours of play) and more time on designing game content. Other players have focused on an album building tool associated with the game, which allows players to make albums containing pictures of their Sims with captions under the pictures. Players have made this album tool over into a graphic novel device and written a great many very long stories about a myriad of themes.

As players have become designers and writers, rather than just players of *The Sims*, they have formed and joined communities on the Internet. These communities, at their best, are true twenty-first-century knowledge-building and learning communities (Gee, 2004; Chapter 6 of this volume). Within them, newcomers are mentored and passions for design, production, and creativity are kindled. The young and old are together in these communities. People, young or old, who choose to put in the practice and time, can become renowned designers or writers with their own large followings.

The Sims communities that have interested us the most, and that we argue hold out the deepest implications for how learning can work in the twenty-first century, are led, for the most part, by females. This is not to say that men are not active participants and leaders as well. But women function in these communities in interesting ways that we will argue betoken where learning for everyone may well go, and may well need to go, if we are to survive in the complex and risky world of the twenty-first century.

We will be talking about gaming among girls and women when, ironically, what they are very often doing is not just playing the game but designing, producing, and creating around the game collaboratively with others across the world. Even when they are playing, they often are also designing new forms of play that in some ways deepen *The Sims* as a game. In Chapter 3, for example, we will see a woman who is designing new, socially conscious ways for people to play *The Sims* and communicate about it.

Gaming as Part of Popular Culture Today

Gaming is going beyond gaming, that is, beyond playing the game to include design, production, creativity, participation, and collaboration in ways that are typical of today's popular culture (Jenkins, 2006; Jenkins, Purushotma, Weigel, Clinton, & Robison, 2009). Video games should not be seen as something separate but rather as a well-integrated part of this productive popular culture.

Many researchers and even the media have been deeply impressed by the ways in which popular culture today recruits digital tools and other devices that foster powerful, deep, and complex thinking and learning out of school (Federation of American Scientists, 2006; Gee, 2003; Johnson, 2006). Let us look at some features of this popular culture.

For many young people today, the digital and the nondigital fully intermix (Gee, 2004). A phenomenon like *Pokémon* or *Yu-Gi-Oh* is not one thing but is represented across a number of different media. *Pokémon* and *Yu-Gi-Oh* are video games, card games played face to face, books, television shows, movies, and a plethora of Internet sites, including fan-fiction-writing sites. Furthermore, in young people's popular culture *Pokémon* and *Yu-Gi-Oh* intermingle with each other, with other similar card games (e.g., *Magic: The Gathering*), and with the anime world more generally. (Anime is Japanese-style animation in books, games, movies, and television; examples include *Pokémon, Naruto, Yu-Gi-Oh*, and *Final Fantasy*.) In popular culture today, diverse media and digital and nondigital technologies converge.

An equally impressive phenomenon has been the ways in which digital tools have allowed everyday people to produce, and not just consume, media (Ito et al., 2008). Today, everyday people can use digital tools to make movies, photos, games, music, newscasts, and many other things. These products can compete with professional work in appearance and, often, in quality.

Connected to this rise of production is a concomitant rise in participation (Jenkins et al., 2006). There are two facets to this rise in participation.

First, people do not need to just serve as spectators for expert filmmakers, game designers, musicians, and news people; now they can readily participate in such activities thanks to the enhanced role of production we just mentioned. More than half of all teenagers have created media content and a third who use the Internet have shared content they produced (Lenhardt & Madden, 2005). In many cases, these teens are actively involved in what Jenkins calls "participatory cultures": "A participatory culture is a culture with relatively low barriers to artistic expression and civic engagement, strong support for creating and sharing one's creations, and some type of informal mentorship whereby what is known by the most experienced is passed along to novices" (Jenkins et al., 2009, p. 3).

Second, thanks to today's digital social networking tools, people can easily and readily form and reform groups to engage in joint activity (such as fan-fiction writing) and even political interventions (such as campaigning) without the sanction and support of formal institutions. Participation today means both participating in producer communities and in many other fluidly formed groups organized around a myriad of interests and passions (Shirky, 2008).

Finally, a number of people working in the area of digital media and learning have pointed out how today's popular cultural activities often involve quite complex language, thinking, and problem solving (Gee, 2004, 2007; Johnson, 2006). The plot of a TV show like *Wired*, with its many subplots and the complex relationships among its characters, is so complex that old television shows pale by comparison (Johnson, 2006). The language on a *Yu-Gi-Oh* card or Web site is more complex, technical, and specialist than the language many young people see in school.

The phenomenon of "modding" video games is a particularly good example of complex thinking and problem solving, often done collaboratively, in today's popular culture. Games today often come with "modding" tools that allow gamers to modify the game by creating new game environments ("maps") and levels for a game, or changing the characters or elements of game play in the game (Hayes, 2008a).

For example, *Tony Hawk* skateboarding games allow players to design new skaters, skateboards, skate parks, and even design new rules and point systems for tricks in the game. Steven Spielberg's game *Boom Blox* allows players to redesign any level of the game or create whole new levels. *Little Big Planet*, on the Sony PlayStation 3, is a platform game where each level is a tacit lesson on how to eventually design your own levels. The game comes with the tools whereby players can eventually make their own games. *Age of Mythology* allows players to create new "scenarios," that is, new landscapes with their own stories and game play.

In most of these cases, players can share their creations with other players. In all these cases, the player is manipulating a three-dimensional (3D) building tool. Such modding is a source of complex technical skills, as well as sociotechnical skills: designers must think about how people will interact with technology, since good game design is focused on the relationship between players and the game.

Skills required to engage in such "modding" look more like the skills that will be important to the twenty-first century than do the skills offered in some of our skill-and-drill test preparation schools (Steinkuehler, 2008a). So do the social, technical, and organizational skills required to lead a guild in the popular massively multiplayer online role-playing game *World of Warcraft* (Brown & Thomas, 2006). It even appears that the reasoning required to engage in debates on many Internet forums involving technical matters (e.g., fans arguing over how the statistical properties determining damage in a game like *World of Warcraft* work) often resembles valued forms of scientific reasoning, forms that we have difficulty eliciting in school with all our textbooks (Steinkuehler, 2008b).

We live in an age of convergent media, production, participation, fluid group formation, and cognitive, social, and linguistic complexity all embedded in contemporary popular culture. Games, most especially *The Sims*, have fully participated in these trends as we will see in this book.

Our discussion of women and games can be seen as one lens focused on today's popular culture. One thing that we will stress, and that has largely been missed in discussions of popular culture, is that the trend toward production and participation is not germane just to the young. We will see women in their sixties exemplifying the trend and even leading some of the young to passion for creativity.

Scholars typically look at things in silos. It is easier to isolate things to study in that way. For example, cognitive scientists study the cognitive processes associated with reading with no regard for the work of social scientists who study the context of actual reading practices. There is also the common separation of presumably cohesive disciplines into highly specialized areas of focus (e.g., the academic study of chemistry has dozens of subfields ranging from organic chemistry to femtochemistry). Thus, it has been common to study games just in terms of game play and to pay less attention to the plethora of things that now go on around games or, at least, not to see gaming communities, modding, and other practices associated with games as all part of gaming today. We will argue that games have become a site for activities and skill building that go well beyond simply playing games and involve crucial, twenty-first-century skills.

By looking at gaming through the lens of how it affects everyday life, we argue that there are very important lessons to be learned about learning in the twenty-first century and even about how we might reform our schools. Before we turn to this topic, however, let us spend some time talking about why, over the last few years, both scholars and policy makers have become so interested in video games. These scholars and policy makers have, by and large, focused on game play rather than the bigger picture we take in this book.

Why the Recent Interest in Video Games?

Nearly everyone today realizes that video games are big business. The vast majority of young people, male and female, have played video games and the average age of gamers is now over thirty-five (Entertainment Software Association, 2009). Video games generate more revenue than do movies, though this does not mean more people buy games than go to movies, since video games are much more expensive than movie tickets (Wolf, 2002, p. 5). Until recently, video games had not really reached a mass audience of the sort television and movies reach. This is quickly changing, as game makers seek to make games that more types of people of all ages can play. This trend has been accelerated by the Nintendo Wii, a game platform whose games have become popular with families playing together and with older people (Daily Mail Reporter, 2009; Douban, 2007).

What many people may find surprising is that video games have recently captured the interest of both educators and policy makers (Federation of American Scientists, 2009; Gee, 2003, 2004; Jenkins et al., 2006; Lenhardt et al., 2008; Salen, 2007; Shaffer, 2007; Squire, 2006). For those who consider games "trivial," this will indeed seem odd. However, video games are built around problem solving in an environment that encourages playfulness and exploration. It is this that has caught the interest of educators and policy makers. In fact, a video game is really nothing but a set of problems or challenges, of different types in different games, to be solved. In a *Grand Theft Auto* game, the problems may be how to pull off an intricate crime spree; in a *Harvest Moon* game, the problems are how to manage a farm and a life in a village. In *SimCity*, the challenges involve planning and running a city; in *The Sims*, the challenges involve building a family and leading a life in a community.

In our twenty-first-century, high-tech world, educators and policy makers have become intensely interested in school reform and in education that can equip young people to live in our ever more complex and risky world (Baker, 2007; Gee, 2008; New Commission on the Skills of the American Workforce, 2007; Pellegrino, Chodowsky, & Glaser, 2001). Problem solving has been front and center in this interest. In the United States and many other countries today, schools have become focused on tests of facts. This focus came from good intentions: the desire to make schools accountable for successfully teaching all students, rich and poor, in a world where, too often, poor students fail in school (Lee, 2006; Madaus, Russell, & Higgins, 2009).

But a focus on such tests has led to two important consequences. One consequence is that even when they can pass information-based tests, many students cannot actually solve problems in the domain where they have passed the test. There is the well-known phenomenon of the "fourth-grade slump" where many kids who pass early reading tests are stumped by the more complex content in areas such as mathematics, science, and social science (Gee, 2008). They have learned to decode print but not to be able to use reading for problem solving and learning in other content areas.

Another example is that many students who pass science tests cannot actually solve problems in science even when the solutions to these problems could be deduced from the answers they have written down on the test (Chi, Feltovich, & Glaser, 1981; Gee, 2004). This lack of problem-solving ability is caused by the focus in school on facts and information, the sort of things that are easy to test, and not on how such facts and information can be used.

Another consequence of our schools' focus on tests is that today's students (not just in the United States but in many other developed countries as well) do not appear to be learning how to innovate and be creative

(Centre for Educational Research and Innovation, 2008; Erlich, 2006; Friedman, 2005; Gee, 2008). Of course, it is hard to innovate in an area when you cannot solve problems in that area and can only write down facts, information, and formulae you cannot apply in any meaningful way. Even when people can solve problems, innovation demands a specific set of skills (Duckworth, Peterson, Matthews, & Kelly, 2007; Dweck, 2006; Lan & Repman, 1995). It demands a mindset where learners challenge themselves, persist past failure, are confident with technical and technological tools, and have a passion for learning. It also requires, more and more, being able to collaborate and share knowledge with others. None of these things—challenge, persistence, confidence with skills and tools, passion, and collaboration—are common in schools today.

Why should we care about problem solving and innovation? There are a number of reasons. One reason is economic (Friedman, 2005). In today's global world, highly advanced technological tools, such as wireless communication, access to the Internet, and high-performance computing, have spread to nearly every country on the globe. Many jobs that used to be performed in developed countries like the United States, whether low status (e.g., a call-center operator) or high status (e.g., computer engineering or radiology), can now be carried out at low cost in countries like China and India. These countries, whose middle-class populations are larger than the entire population of the United States, even leaving aside the massive number of poor people in these countries, can use the Internet and other technologies to perform jobs for businesses and institutions in developed countries. Apart from jobs that demand physical face-to-face encounters, the job market has now become global, and U. S. workers often compete with workers around the world.

This means that many U. S. workers, and workers in other developed countries, who have only standard or standardized skills are in peril. They will have to work for less financial gain or watch their jobs go overseas. For big rewards in a developed country, people must be able to go beyond standardized skills and be able to innovate. Developed countries, to compete, will have to become centers of innovation and creativity. This is not likely to happen if the U.S. school system stresses standardization, basic skills, and passing fact-based tests and not problem solving, innovation, and deep thinking.

Another reason we should care about problem solving and innovation is historical. The world has changed. First, thanks to the Internet, as well as many other knowledge-building technologies, facts and information—what so often counts as "content" in school—are "cheap." Students can readily look up on the Internet anything a teacher is telling them. Indeed, they can find out things that the teacher does not even know. In such a world, just having facts or information are not twenty-first-century skills. Rather the twenty-first-century skills related to information are how to assess information, how to filter it for successful use so as not be overwhelmed by it, and how to use it for solving problems and innovation. These abilities are rarely tested in our schools today.

Second, thanks to the rapid growth of science and technology, jobs, skills, and information readily go out of date. No one can trust that skills learned at one point in life will not be obsolete later. People of all ages will have to know how to learn new skills, and not just superficially (remember the worldwide competition), over and over again through a lifetime. This will require being open to challenges, persistence, passion, and often collaboration. Each new learning opportunity will require problem solving and the ability to innovate if one does not want to be "left behind" (to use the current educational slogan) in terms of opportunities for career advancement and to remain an active, informed, and involved citizen in general. Twenty-first-century schooling must become focused on preparation for future learning.

Yet another reason we should care about problem solving and innovation is social. Thanks to globalization and the intensive international competition, in developed countries most people will not be able to get jobs whose high pay and prestige bring them status and power in society. It has been argued that only one fifth of the workers in a developed economy will be highly rewarded, and these rewards will be for knowledge building, problem solving, and innovating new ideas, products, services, technologies, and structures (Reich, 1992, 2001). The remaining workers will be much more poorly rewarded service workers (the biggest category of workers in a developed country) and the few remaining industrial and manual workers in developed economies.

Not all people will be able to gain status and a sense of worth in the market in terms of the money they make and the prestige of their jobs. More and more people are turning to activities "off market," or ones they

do not do for money, to gain status and a sense of belonging. They are joining communities on the Internet, for example, to build and design things
and to help others (Anderson, 2006; Leadbeater & Miller, 2004; Toffler &
Toffler, 2006).

In this book we will discuss elderly women who, late in life, have become
experts at designing clothes, furniture, and houses for a video game, *The
Sims*. They engage with a community wherein they mentor newcomers and
gain a good deal of status and adulation, none of it for money. We will see
a young girl doing poorly at school but becoming a skilled designer replete
with high-tech skills and a large audience of fans. We will see a graduate
student who gained a graduate education before going to graduate school
as a designer and leader "off market" in *The Sims* and in *Second Life*.

Today, there are such communities on the Internet and in "real life" giving people status "off market" and built around almost any interest you can
think of. It is crucial for our future that schools see to it that all learners
pick up skills and passions, and learn how to gain new ones in the future,
that allow them to gain a sense of worth and to contribute to others, even
"off market." Otherwise, we will continue to face a great divide between the
rich and powerful, on the one hand, and service workers, on the other.

What has all this got to do with video games? Video games are virtual
worlds in which players, individually or collaboratively, solve problems in
highly motivating contexts (i.e., in interesting worlds or as parts of interesting stories). They do so in an environment where the cost of failure is
low enough that exploration, risk taking, and trying new things is encouraged. Further, there is often a premium on solving problems in games in
different and novel ways (Gee, 2003, 2004, 2007).

Good game designers seek to free players to try different things, to solve
problems in their own ways, to engage in their own styles of play and learning, and to try out new ones. We can make games about problems in, say,
urban planning, civics, or engineering, as well as we can about shooting and
killing. Indeed, today there is a robust independent game industry and a
so-called serious game industry that are making games that go well beyond
shooting and killing in a variety of different directions, including games
for health, social change, and learning content relevant to school (see, for
example, http://www.gamesforhealth.org, http://www.gamesforchange.org,
and http://www.seriousgamessource.com).

We have both written a good deal about games and learning, espousing the sorts of problem-based learning that games can excel at (and such learning does not require games but can be accomplished in many different ways). Today. there is a great deal of literature on games and learning. As we said in the last section, we are concerned in this book with looking at games' contribution to twenty-first-century learning in a somewhat broader fashion. In today's games, the players become themselves experts in the problem domains the games are about.

Emotional Intelligence in a Complex World

Though there is an intense interest in school reform and twenty-first-century skills in the United States (see Partnership for 21st Century Skills, 2004, 2007), too often this interest focuses narrowly on so-called STEM—science, technology, engineering, and mathematics (see, for example, STEM Education Coalition, 2009). U. S. students do not fare particularly well on international tests of science and mathematics (Lemke et al., 2004; Programme for International Student Assessment [PISA], 2006). Since many believe that science and technology are the keys to success in the twenty-first century, school reformers and policy makers have become obsessed with the STEM areas.

There is, however, a deep problem with this STEM obsession. We live in a world that is, in Thomas Friedman's (2008) words, "hot" (global warming), "flat" (global with everyone competing with and influencing everyone else across the globe), and "crowded" (overpopulated and with massive numbers of poor and starving people). As we write this, the world is in the clutches of a global economic meltdown, the causes of which appear to be too complex to be readily understood by standard economists. Many economic experts have admitted that they never saw it coming and nothing in their training and experience prepared them for it (see Kiel, 2008 for a video of Alan Greenspan, a man crucial for managing the U.S. economy for decades, admitting he does not understand the economic crisis).

We are all well aware that the world is dangerously awash with cultural and religious strife that often masks economic conflicts and the divide between the rich and the poor in the global world. Global warming, agreed upon by all scientists but still treated as only a possibility by our media and denied by many on the right wing, is already wreaking havoc through storms and shrinking coastlines but will eventually bring environmental and economic disaster to the world if not stopped (Brown, 2008; Gore, 2006).

The unintended consequences of our scientific, technological, and political decisions in our global world, a world replete with interacting complex systems, are an ever-present danger. Today human and the natural systems continually interact in complicated ways. For example, consider that, for most of history, weather and storms were an "acts of God," but now, thanks to human industry, pollution, and interventions of the sort causing global warming, they are also "acts of humans" (Brown, 2008). We live in Phoenix, Arizona, where the endless concrete parking lots, roads, and strip malls create a heat island effect (the man-made environment radiates heat at night that has been built up and stored all day) that does not allow the temperature to cool at night, as it normally does (and as it does nicely in more open lands of northern Arizona where we also live). Thus, the high temperature in Phoenix is partly human made.

In such a world, with its complex and interacting social, institutional, cultural, religious, and economic forces, no single area of expertise can solve our problems. Today's problems are so complex that they require the concerted and collaborative efforts of people with many different areas of expertise. They also require the humility of experts, who must to acknowledge the limits of their expertise.

Problems in the modern world cannot be labeled as purely scientific, purely social, or purely political. The scientific, social, and political mix together and with other factors (such as economics, culture, religion, and so forth). Finally, and most importantly, in order to solve problems today, we often need to change not only policies but also people and how they think and behave. This could not be clearer when we think about global warming, but it is true of so much else today (e.g., providing health care to the many people who are not covered by insurance may well require that people with more resources and good insurance accept that decisions about who gets expensive treatments will be made on the basis of equity and possible lifespan and not privilege—not an easy sell when you are one of the privileged).

In such a world, the scientific and the technological must be integrated with what has been called "emotional intelligence" (Goleman, 1995). Unfortunately, this phrase has become trendy and has spawned a plethora of definitions and confusion. What we mean by the phrase is simply this: emotional intelligence (which we argue can be learned and developed) is the ability to understand and manage one's own emotions, decisions, and perspectives, as well as to understand and helpfully change or collaborate with other people's emotions, decisions, and perspectives to make the world a better place. This is not how the term is officially defined, but it is how we will use it. Note we are only interested in emotional intelligence put to positive uses (if you like, call it "positive emotional intelligence"). We are not interested in "evil geniuses" capable of manipulating people's emotions to bad purposes.

People's decisions and perspectives are in our definition of "emotional" because how people look at the world and the decisions they make are not detached from their feelings and emotions. In fact, it is feeling and emotion that make some things more appealing to a person than others, and this sense of what is appealing helps guide perspectives and decisions. People have an "appreciative system" (Schön, 1983) in which they judge whether the outcomes of actions they have taken are "good" for their goals or not. If their actions are not, they act again. Without such an appreciative system, which is based not just on rational calculations but on "how things feel" and whether they "feel" right or not, people could not make decisions. They would be at a loss amid all the possibilities. At the same time, we can change our emotional responses and what feels right or not, especially if we interact and learn from others.

Emotional intelligence, as we have defined it, is, in our view, closely related to social intelligence (Goleman, 1997): the ability to understand and manage social relationships and to collaborate with others. Human beings' emotions are almost always caught up in, determined by, and influence their social relationships. Our social relationships, from our early socialization in the family to our adult relationships, are formative for our emotional lives. We humans are social creatures. We really should talk about a socioemotional intelligence.

In a complex world, full of risk and unintended consequences, it is important that scientific, technical, and technological thinking and decisions—forms of cognitive or rational intelligence—be fully integrated with socioemotional intelligence. To solve our problems, people, and not just things (e.g., nature, the environment), need to be variables in our solutions. People are the sort of variables that have emotions and relationships.

It is not just minds but emotions and relationships, too, that must change if real solutions are to take hold.

It is not just socioemotional intelligence that must be integrated with STEM skills for a complex, risky world. We have also, in our obsession with STEM, created a great divide between art, on the one side, and science and technology, on the other. In fact, in our rush for STEM funding, we researchers have by and large ignored the arts. This is a profound mistake and quite dangerous. First of all, the arts are one source of creative, innovative, ethical, and exploratory thinking that can not only bring socioemotional intelligence to our problems but also can drive us to see things in new ways leading to new solutions.

Second of all, today there is no real divide between technology and art. Artists today, whether engaged in music, photography, graphic art, game design, or nearly anything else, use cutting-edge digital technologies in their work (and we will see this throughout this book). Photographers use Photoshop to alter their pictures and graphic artists draw on computers, not paper. Artists today are as adept at digital technologies as scientists are.

Indeed, we have a colleague at Arizona State University who is making cutting-edge collaborative environments where people interact face-to-face and simultaneously with 3D digital images to learn in the STEM areas (see K-12 Embodied and Mediated Learning, 2009). He is, by training, a musician, though one that used a great deal of technology to produce music. There is no reason today why science and art students are not together in a college classroom using the same digital tools (e.g., simulation software) for discovery both in the scientific sense and the artistic sense and for communicating key ideas to the public.

So, while it is certainly against the current trend, we argue for educational efforts that integrate STEM, socioemotional intelligence, and art. STEM left isolated will not solve our problems. Indeed, it may make them worse.

Back to Gaming Beyond Gaming

In this book we will see girls and women gaming and going beyond gaming to design, produce, write, create, and collaborate. They will be solving problems as highly proactive learners. They will be participating in learning communities with very special features that help people of all ages and backgrounds to build passions and gain expertise. And they will be doing all this by combining very real high-tech skills with socioemotional intelligence and artistic sensibilities. We argue that they are a better guide to where we need to go to enhance twenty-first-century learning, in and out of school, for everyone than are the many purely technicist forms of school reform today.

We will focus on the stories of individual girls and women, but always with an eye on big issues about learning in our complex, globalized world. Some people have considered girls and women as marginal to gaming—most certainly the elderly women we will meet—but we argue that they are central to where gaming, popular culture, and learning is going in the future. Games are going beyond gaming, and while women are not the only force taking them there, they are fully part of the action. They can help us greatly broaden the discussion of learning in and out of school in the twenty-first century.

Chapter 2

Video Games and Twenty-First-Century Skills

Why the Sudden Worldwide Interest in Video Games and Learning?

Age of Mythology

This book is about girls and women gaming and taking gaming beyond simply playing a game to other game-based activities that we argue represent new and powerful forms of twenty-first-century learning. But before we move on to our main focus, let us start with the recent worldwide interest in video games and what they have to teach us about learning in the modern world. Why have video games become so interesting to educators and policy makers? This chapter will seek to answer that question. We also hope to give readers a bit of background on video games and learning to prepare them for the rest of the book.

Imagine someone offering to transport you "to a time when heroes did battle with monsters of legend and the gods intervened in the affairs of mortal men" (Microsoft, 2002). You will actually enter and act in this world; you will not just watch it. This is just what the video game *Age of Mythology* (first released in 2002) does. You take on the role of an ancient civilization—the ancient Greeks, Egyptians, or Norse—and command all aspects of an empire that you develop from its earliest origins to its days of greatness. You set your people to work gathering resources and constructing buildings, settlements, and cities. You raise massive armies and wage Homeric-style wars against your enemies. You establish trade routes, explore new frontiers, and expand your empire as you advance through the ages, all the while you call on ancient gods and mythological heroes to aid you. In fact, as you look down on all your people and your growing empire, you are yourself a kind of god, calling into existence an ancient civilization shaped to your desires and whims.

Age of Mythology, often referred to as *AOM*, is a real-time strategy game. Real-time strategy (RTS) games are among the most complex video games available, involving the mastery of hundreds of different commands. In RTS games, games such as *Age of Empires, Rise of Nations*, and *AOM*, the player develops a whole civilization from its earliest primitive days to its full development as a large civilization or empire. The player sets his or her people to gather resources (things like wood, stone, minerals, gold, oil, and so forth) and uses these resources to construct buildings of many different types within settlements and cities, and to earn more citizens who can play different roles (e.g., as workers, scouts, soldiers, leaders, and scholars). Resources are also used to construct and train different sorts of military units and to build a variety of weapons and technologies.

In RTS games, the player manages every aspect of his or her civilization, including warfare, diplomacy, taxation, religion, territorial expansion, and other elements. As the player's civilization develops and the player accomplishes certain goals, the player earns the right to move to a new, more modern age. Different RTS games move players through a variety of different ages, and in each new age the player gains new powers, skills, and technologies for use by the civilization's citizens and soldiers, as well as upgrades and improvements to building, settlements, and cities. *AOM* has four ages: the player starts in the "Archaic Age" and moves through the "Classical Age" and the "Heroic Age" to the "Mythic Age." In a game like *Rise of Nations*, the player moves from primitive times to the nuclear age through a wide variety of different ages.

In its initial version, *AOM* had three playable civilizations: the Greeks, the Egyptians, and the Norse. The player had to choose one major god for his or her civilization before beginning the game. Each civilization had three possible major deities from which the player could choose: Zeus, Hades, and Poseidon for the Greeks; Isis, Ra, and Set for the Egyptians; and Thor, Odin, and Loki for the Norse. Each time the player advanced to a new age, the player could select a "minor god," gods like Bast and Aphrodite.

All gods, major or minor, grant to the player unique technologies, special mythic units for warfare, and a unique god power, a special ability used to help the player or damage opponents.

An expansion to *Age of Mythology* (a sequel building on the previous game and requiring the player to own the previous game) was released in 2003, called *Age of Mythology: The Titans*. This expansion added a new civilization, the Atlanteans, as well as several new units, including the Titans, powerful, gargantuan, godlike beings that players could summon to their aid.

What makes RTS games "real time" is that events are happening simultaneously. While the player is making and implementing the myriad of decisions needed to build his or her civilization, other players are doing the same thing for other competing civilizations at the same time. These other players can be real people when the game is being played as a multiplayer game, in which several different real people play against each other from different, but linked, computers, or the other players can be computer-controlled players—with decisions enacted by the computer—when the game is being played as a single-player game. In this case, the player is pitted against other civilizations controlled not by other real players, but by the artificial intelligence of the computer software. Winning the game does not always require military conquest; in some RTS games, players may use diplomacy to achieve global leadership.

Since the game is played in real time, the issue of speed arises. As you make and implement decisions, the game is not paused and other players are making decisions and taking action at the same time. One strategy in RTS games, called "rushing," is to build up your civilization's military units as quickly as possible, typically by neglecting other aspects of your civilization. You can then use your armies to rush your opponents, possibly catching them at an earlier and less powerful stage of development than your civilization. Because speed can be important, advanced players often remap the commands in an RTS game to be faster and more efficient, choosing, for example, to map a series of key strokes that carry out several related commands to one "master" key stroke that signals to the computer to do the whole series with one command.

In contrast, there is another strategy called "turtling" (being a "turtle," slow but well defended). This strategy consists of building strong defenses and accumulating many resources, so that when another player comes to attack, you will be too well developed and too strong to be easily defeated.

When you are well prepared and ready, you set out with a powerful and sophisticated army to conquer your opponents. Of course, if someone who is turtling is rushed by other players too soon, they can take heavy loses, just as a player who rushes too late against a well-built, turtling opponent can take heavy loses. Different RTS games make rushing or turtling more or less viable strategies, though players can often set conditions in their games that allow or encourage rushing or turtling.

Let us return to the offer to be "transported to a time when heroes did battle with monsters of legend and the gods intervened in the affairs of mortal men." This sounded great to start with, but now it also sounds difficult and complex. A game like *AOM* is, indeed, demanding, and our description left out a great deal of complexity so as not to overwhelm readers unfamiliar with video games. If we can overwhelm people who are just reading about such games, imagine how overwhelming they might seem to new players learning to play them for the first time. Being god is hard work, with much to think about and do and very little time for either.

AOM is played, like many other video games, by children and adults. Oddly, many adults are more intimidated by such games than are young people. Knowing that they can be god, but faced with a big learning curve, more seven-year-olds will take the offer than adults will. Let us tell you a story about one such seven-year-old.

Sam's Story

Today, Sam is fourteen, but our story happens when he was seven. *AOM* was one of the first highly sophisticated games Sam played, though he had interacted with computers and children's games since he was two. One day, when Sam was in second grade, he asked his father if he could get the game *Age of Mythology* and play it. Sam's father told him that the game would be too hard for him and that he was too young to play it. Sam replied that several kids in his second-grade class played the game. His father responded by saying if that was true, they must be playing it with their parents' help. Sam was certain that they played it on their own. His father relented and bought Sam a copy of *AOM*, intending to learn to play it himself and help Sam learn it.

Things did not turn out as planned. The father, in his fifties at the time, found the game difficult and learned to play it more slowly than did Sam (who eventually helped his father catch up). Sam played the game with great enjoyment on his own, as a single player game against civilizations controlled by the computer. Typical of many parents, Sam's parents did not pay close attention to how he played once they knew he was comfortable with the game.

One day, Sam's father was bragging to an adult gamer that his seven-year-old played such a complicated game as *AOM*. The gamer said that winning the game required rushing, not turtling. The father became concerned, since he knew well that Sam loved to build things, but did not like time pressure in anything he did. He began to wonder whether Sam ever won the game, and why he liked to play it.

So Sam's father asked him if he ever won. Sam said he always did. His father asked Sam if he could watch him play the game, something Sam, like many other young gamers, was more than happy to allow. And sure enough, Sam turtled, building quite methodically and enjoying the ever more complex growth of his original buildings and settlements.

His father wondered why no other civilization rushed Sam and caught him unprepared for a quick battle to the death. But then the father saw a purple hippo with a top hat fly across the screen. The father was surprised, to say the least. He said to Sam, "I know my knowledge of ancient mythology is rusty, but I don't remember any purple hippos." Sam replied that he had added the hippo himself, using a device called "a cheat," a little bit of code that can be typed into the game to add or modify game features.

The father asked, "So, then, Sam, what does the hippo do?" as another, and then yet another purple hippo flew by. Sam moved the image on the screen away from his settlement to the base of an enemy civilization. Purple hippos were spraying purple paint on all of the buildings as the other civilization's workers built them, slowing down construction as they stopped to clean the buildings and preventing the possibility the civilization would develop quickly and rush him. By the time Sam had fully developed his original settlement, his armies attacked with great success.

Sam did what lots of gamers do. He refused to accept the game the way it came. He wanted to learn it and play it his own way. He wanted to enjoy building, to turtle and still be victorious. He discovered the cheat, the way most gamers do, by talking with other gamers and surfing the Web to get information about his game. Sam turned *AOM* into a building game, friendly to turtling. In this way, he spent hours learning to build and develop his civilization without lots of interruptions to fend off early enemy attacks. What Sam did, and we will talk more about this later, was a form of what we will call (and gamers call) "modding." Modding refers to how gamers modify their games, or redesign them, for various purposes, including to make the game support their desired style of play.

Sam's father was impressed, but soon worried again. When he told the purple hippo story to a friend at a professional conference, the friend said that Sam was using a "crutch" and thus not really learning to play the game. The father was crestfallen. He called Sam on the phone and told him that the purple hippos were a crutch, keeping him from really learning the game.

Sam said, "Don't worry, Dad, I'm down to a one-hippo game." Eventually, Sam used no hippos, rushed and dealt with rushes, and became an expert *AOM* player. He has never regretted his use of purple hippos. Sam was customizing *AOM* not only to his playing style, and to his delight in building, but also to his learning style and dislike of time pressure. Sam was very much in charge of his own learning and saw to it that his game play resulted in mastery on his own terms.

Building and Designing

Just because Sam eventually learned to play *AOM* without purple hippos does not mean he gave up his love of building. He was, of course, now a good enough player to build a good deal without hippos, since he was now adept at fending off early rushes and even could engage in some himself. But he also discovered that he could mod the game in much deeper ways than adding purple hippos—in ways that involved a good deal of building and design.

AOM comes with something called a "map editor" or "scenario editor." This is a tool that players can use to create their own games and let other players play them. Such player-designed games are called "maps" or "scenarios." Players use the same sort of software used by *AOM*'s professional game designers to build their own physical environments (lakes, mountains, valleys, and so forth), buildings, towns, and characters, as well as narrative and game-play elements. They also can modify the artificial intelligence of any characters the computer will control in the game. For example, one player designed a scenario called "Stranded Romans" about a group of Romans stranded on an island who have to fend off the native population of the island. The player then uploaded this game to a fan site, *Age of Mythology Heaven* (http://aom.heavengames.com/), and other players downloaded it and played it, eventually giving him feedback about what they thought of his game.

Another player created a scenario called "The Lost Legion" about a legion that vanished in the forests of Gaul. In this scenario, in the words of a reviewer, players battle "across this 1000×1000 map that is completely jam-packed with stunning eye candy like multi-story buildings, ports, cities and magical outposts as you follow the incredible story of Marcus Scarus! The map is filled with twists and turns and guarantees at least 2 hours of fun" (The Lost Legion, 2004).

Players do not have to build games with the scenario editor. They can just build environments they like. For example, one player made a map of the Netherlands just because he wanted an accurate map of his country's environment that others could use to create game scenarios. A player might design a mysterious swamp or an ancient city. Players can even build *AOM* environments, place characters in them, and simply watch while the computer controls the action, like a movie.

The scenario editor allowed Sam to build and design for hours. He did not build games for other players. Rather, he enjoyed building rich worlds where mythological creatures interacted in marvelous landscapes. The game as such had disappeared. Now he was involved in a new form of "writing," or "drawing" if you like, using digital tools to make worlds and imagine the stories that happened inside them. Other players use the same capacity to design full-blown games for themselves and others, becoming game designers in their own right—and some eventually enter game design as a profession. But for Sam, creating the scenario was like writing or drawing, albeit with digital tools.

In fact, Sam made deep connections between games, most certainly including *AOM*, and reading and writing. He did not see them as separate or competing activities. When Sam grew passionate about a game, as he had about *AOM*, he wanted to read and write about it and eventually connect it to other activities and interests.

Reading and Writing

Sam and some of the other second graders in his class were inspired by *AOM* to develop a passion for mythology. They regularly went to Internet sites to learn more about the game and about the mythology connected to the game. They started to write their own stories about characters in the game and in mythology more broadly, some of which they later would upload as fan fiction to the Internet. Their interest in mythology extended well beyond the game, as they searched their school library for books on mythology, some of them quite old and dusty. They shared their stories and the books they found with each other, widening their reading and writing to mythologies not in the game, such as the myths of American Indians.

As is common, these children's parents (all these parents were middle or upper-middle class) cultivated their children's growing passion for reading, writing, and drawing about mythology. They bought their children more books, including books primarily meant for older children and adults. As the children gobbled up all they could find in books and on the Internet about mythology, their reading level for such material skyrocketed. Their parents took them to museum exhibits and any other events that could be related to mythology, such as American Indian cultural events. On trips abroad, the parents went out of their way, with their children's encouragement, to investigate the mythological stories of the places they visited.

The parents encouraged their children to teach them about mythology and to discuss what they read, wrote, and learned. They encouraged their children to be experts and treated them like experts. They encouraged them, as well, to relate their passion for mythology to other things, such as natural history, cultural differences, history, and languages. In a term we will discuss later, these parents were "cultivating" their children (Lareau, 2003), purposefully guiding them to become the sorts of self-motivated and confident "knowers," readers, and writers that are successful in school.

Today, Sam no longer plays *AOM*, having moved on to other games and other activities that were inspired by his game play (such as *Dungeons and Dragons*, a role playing game played face to face, that stresses the players' own imaginations, and acting in children's theater, an activity that Sam saw as a yet more direct participation in imaginative worlds). But *AOM* and the passion for mythology it ignited became part of the furniture of his mind and imagination and part of the foundation for his higher-order literacy skills.

Tech-Savvy Skills in a Social Setting

AOM also became, for Sam, a gateway to tech-savvy skills. He had been comfortable with computers since he was two. However, *AOM* taught him about modding, as he did when he used the purple-hippo cheat or used the scenario editor. As for many other children, especially boys, modding allowed Sam to becoming proactive with computer technology. Sam became confident in his interactions with digital tools. He saw them as means to produce meanings and emotions, just as writing and art do.

As he played more games, Sam modded more and more, and also learned to make Web sites devoted to his favorite games and characters. As he played games on game platforms like the PlayStation 2, he modified the hardware of his PlayStation so it would play Japanese games because he had developed a passion for anime and anime games.

Using technology to produce and not just consume was for Sam, as for so many other children, tied to participating socially with others in person and virtually. Any time he became passionate about a game, he joined forums to discuss it and sometimes constructed Web sites to organize discussion about the games, their characters, and their stories. He also regularly engaged in face-to-face interactions with his friends while gaming together in the same room. He shared, as well, his extensions of gaming into writing and reading, as he did when he shared his growing passion for reading about mythology with his friends who played *AOM* or shared stories he had written inspired by *AOM*. For Sam, and nearly all the children we have studied since we started our studies of games and gamers, gaming is a social experience, not isolating (see Lenhart et al., 2008 for more information about the social nature of gaming).

Sam, like many other children, developed an attitude toward digital technologies as being about production, participation, and authorship, not passive consumption. At the same time, Sam did not see technology as divorced from literacy (e.g., reading and writing) or art (e.g., drawing or music). As he got older, he was as likely to use digital tools to engage with art as he was with science, in and out of school. Further, he did not see digital technology as isolated from things like face-to-face role playing in *Dungeons and Dragons* or acting in theater productions. All of these were connected in his life, just as his interest in *AOM* was connected to so many other activities. Games were not something just to "waste time" on, unrelated to more "important" matters, no more than were books.

Video Games: Are They Good or Bad?

Video games have become big business. They are, in many ways, the preferred media for many young and not-so-young people, since the average age of gamers is now around thirty-five (Entertainment Software Association, 2009). The gamer generation is getting old. As games have become pervasive in our society—indeed, across the whole world, since it is hard to find a place where people do not play video games—they have also become controversial.

In some ways this controversy is just typical of new technologies. Writing was controversial when it began. Plato, for example, thought writing would kill memory and allow people to claim to know what they had not really mastered when they committed it to memory (see Plato's dialogue entitled *Phaedrus*; for explication of Plato's critique of writing, see Gee, 1990; chapter 7). Print was controversial when it began to be used. A printing press made books easy to produce, since it was much easier and cheaper to print a book than to write it out by hand (as was done before the advent of print). This ease of production gave rise to fears that many trivial books would be produced and people would waste time reading silly things. There was also the problem that the mass production of books made it easier for working-class people to get access to them. The working class might then read and even write materials that would make them want to rebel against their "betters"(Altick, 1957).

Of course, violence in video games has been foremost in the controversy around games. Though thousands of video games have been produced and sold, media coverage at times would have one think that *Grand Theft Auto* was the only game ever made. Many games are not violent, including the game that will eventually be the focus of this book, *The Sims*. Nonetheless, many very good and respected video games, like *Deus Ex*, *Half-Life*, *Metal Gear Solid*, and *Call of Duty*, are indeed violent. There are several points we want to make about violence in video games, in part so we can move on from this riveting topic to the matters we want to discuss.

First, there is quite a bit of research on violence and video games, and overall this research does not yet lead to any definitive conclusions (Kutner & Olson, 2008). It is clear that violent games excite people (particularly young boys) for a while after they play such games, though it is quite unclear that, save for already violent and disturbed people, these games prompt people to engage in violence. In fact, aggression and violence are caused by people living in a culture or environment of aggression and violence that extends far beyond video games (Dowd, Singer, & Wilson, 2006).

The same can be said for books with controversial or violent subject matter. Books are, indeed, very powerful, and we would argue that they still are far more influential than video games. We readily admit we do not want young children playing mature-rated games or people already prone to aggression and violence playing violent games (just as we do not want them to read violent books).

Second, well-designed and extensive research shows that human beings treat what they see on screens, such as movies and games, as "real" in the sense that they respond to it emotionally as if it were happening in the material world (Reeves & Nass, 1999). This is probably the case because evolution never prepared the human mind for differentiating emotionally between seeing two people kiss on screen and in the real world, since screens were nonexistent as our minds evolved. This "confusion" between screens and life is, of course, a considerable part of media's power. At the same time, it is also an effect we choose to accept and that we can stop by using our higher mental faculties—our critical thinking skills—by saying to ourselves, "This is just on a screen." All mentally healthy people—as well as some mentally unhealthy people—know the difference between screens (and fiction generally) and reality. People who do not know the difference are missing critical faculties and are dangerous to themselves and others in all sorts of ways, well beyond what video games they play.

Third, and most important, video games, violent or not, are not simply good or bad. Our media always demand black and white, unequivocal answers and categories: things are good or bad in and of themselves. Technologies (including books, which are, indeed, technologies for meaning making) are never in and of themselves good or bad. Their effects depend on how they are used and who is using them to what ends. It is a matter of context. Video games and books can be used for good or ill. They can be a productive use of time or a waste of time. It all depends on what we do with them.

Sam was most certainly not wasting his time playing *AOM*, any more than he was wasting time when he was reading or writing about mythology. On the other hand, if a child is engaged with any activity—games, books, or television—that is not being thought about, connected to other things, and made part of social and intellectual interactions with others, then the activity may be less beneficial, though we still have to consider the situation carefully, since even activities that may not seem "productive," like relaxation and day dreaming, can be valuable in certain contexts. The outcomes of gaming are not black or white and are not simple. We actually have to consider in each case what the child is doing with the game, as well as what parents and peers are doing with the child. We have to ask not just what the child could do differently but also what we could do to help and mentor the child to use the game in as fruitful a way as Sam and his friends used *AOM*.

The Moral of Sam's Story

Now back to Sam and what his story has to tell us about how to make games better for people and society. Many educators and policy makers today have become interested in the potential video games hold out for reforming learning and innovation in schools and society (see, for example, Federation of American Scientists, 2006; Gee, 2003; Salen, 2007, among others). Some of the key themes in Sam's story show this potential:

1. Video games led Sam to become a proactive producer with digital technology. They also served as a gateway to the mastery of other digital technologies.
2. Video games led Sam to develop higher-order literacy skills as he read and wrote more about mythology. Furthermore, much of the language he saw on Internet sites about the game and in the reading he did about mythology was complex and demanding. He saw varieties of language that were technical, specialist, and academic, language officially beyond the reading level of a second grader but language he could handle

because he had built up such passion for and background knowledge in mythology.

3. Video games led Sam to learn how to use digital tools socially, how to participate in Internet communities, and sometimes even help form them and lead them himself. Sometimes he taught or mentored others, other times he learned and followed. He learned to engage in collaborations on the Internet and in the real world in order to play and design in regard to *AOM* and his other game interests.

4. Video games led Sam to a deeper interest in the real world. *AOM* made mythology a passion for Sam and led him to study mythology in the real world.

5. Video games motivated Sam to combine and integrate reading, writing, art, content learning (around mythology), and real-world exploration (like his trip to Peru).

6. Video games taught Sam to think and act like a designer. Thinking like a designer, including understanding how the parts of systems work together to achieve desired functions and goals, is an important skill in the twenty-first century as we confront so many human-made, natural, and blended (both natural and created or affected by humans) systems like the global economy, the environment, or the clashes of civilizations and cultures.

7. Video games allowed Sam to interact with and get mentored by a large number of older peers and adults as he engaged in what was not an age-graded activity. His parents cultivated his interests, but he was also helped and mentored by more advanced peers and adults in the various Internet communities in which he interacted. When he later went into acting, he was adept at interacting with older peers and adults.

Surely all these experiences would be good for all children to have. Sam's gaming not only did not get in the way of his success in school but also enhanced it by building thinking, learning, and literacy skills. His gaming supplemented his schooling, developing his skills at producing, participating, collaborating, innovating, creating, and designing, skills that were not readily developed through the curricula in his school or many others. It has been argued, and we will argue the case in this book as well, that these are important twenty-first-century skills for our world (Partnership for 21st Century Skills, 2007).

Games and Learning

Sam's story makes it clear, at least partially, why so many educators and policy makers (at least those who can move beyond violence in games) see the potential in video games to enhance and reform learning in schools and society. In fact, a whole movement has formed around this interest over the last few years (see, for example, Federation of American Scientists, 2006; Klopfer, Osterweil, & Salen, 2009).

But now we must stop a moment and say that we misled you in certain ways. We picked a game that has "content," namely mythology, that is sufficiently associated with school such that many readers will readily agree that learning such content is valuable. We also picked a child from a well-off, though not highly wealthy, family. Sam's parents were both professionals. The issues of game content and socioeconomic status of game players are important when we consider the future role games can play at school and in society.

Let us start with game content and take up later the significance of family economic issues. Most video games do not have content that is in any way reminiscent of content in school, content like algebra and civics (though people have designed so-called serious games to teach algebra and civics and many other such things, and continue to do so today). On the other hand, we will argue that content in school is part of the problem and not the solution to reforming our schools, as odd as that may sound.

As an example, we will use a worst-case scenario, the notorious *Grand Theft Auto*, or *GTA* for short. We pointed out that many games are not violent, but some games are, and it is important to make our point with a challenging and notorious case. There is an interesting paradox with *GTA*. People in the media and politics tend to hate it, decrying its violence and some of the immoral things players can do in the game if they choose (like going to a prostitute or killing innocent people or policemen). Gamers, on the other hand, revere the game, but not because it is violent or immoral.

GTA was not designed for children (as we mentioned earlier, all of the *GTA* games have an M rating and thus are designated as appropriate only for players at least seventeen or older), and they should not be playing it. But the older teens and young adults (and not-so-young adults like these authors) who play the game revere it as a superbly designed game that turns the player loose in an extensive world with lots of choices to be made and lots of places to be explored. Similar to many game series we discuss in this book, there are several *GTA* games, since sequels to the original game have been made over the years. Each sequel explores a new world and a new story. In all cases, you play a character who is poor, needs to make money, and must find his way in the world (all of the GTA protagonists are male). Engaging in petty or major crime is one way you can proceed, though you can also drive an ambulance or a taxi cab. You can be respectful of little old ladies on the street or kill them.

The first intriguing feature of *GTA* games is that the player can make moral choices, choosing how bad or good to be. Of course, if we removed all bad choices from a game world or the real world, there would be no choice or moral universe left. The second and yet more important thing is that mature gamers love *GTA* games not because of their violence but because the games offer a continuous series of problems that take skill and intelligence to solve. Players are continuously making choices that they have to live with and whose consequences they have to deal with.

The content of many video games *is* the problems to be solved. Video games are all about problem solving and the sorts of persistence, ability to cope with failure, and strategic thinking that problem solving requires. The story and images in the game motivate the player's desire to be in the game's world, solving the game's problems.

Imagine we changed *GTA* so that players never shot people, they just took pictures of them. Instead of, say, planting a bomb in someone's car and killing them, the challenge was to plant flowers in the car without being caught and take a picture of the person surprised by the flowers. The challenge would be the same; the basic problem-solving design of the game would not have changed greatly. The game would not sell as well, since the popular-culture, media-driven themes that *GTA* plays on so well would be largely missing. But the problem solving that drives players to mastery in the game, and the key thing that keeps them engaged hour after hour, would still be there. After all, gamers persist for many hours solving problems in games like *Super Mario Brothers*, where there is no real violence.

A game like *GTA* uses problem solving, challenge, story, choices, and open-ended exploration so well that it inspires educators to ask whether we could not do the same thing with different content. Why not use the world and problems of astronauts, city planners, activists, or scientists to the same effect? In fact, people are making games in all these areas, because they see certain forms of problem solving as the main event when it comes to learning, not content in and of itself.

If we think about content in school, what does it really mean to learn algebra or civics or anything else that we count as a subject in school? What do we really mean when we talk about learning "content" in school, or when we bemoan the fact that U.S. children do not do as well as their Chinese counterparts on international comparisons of mathematics and science knowledge? When we say that video games do not have real "content" in the way in which school does? Very often—too often—in school, what we mean by content is a rather inert body of facts and information. Algebra or civics is treated as just a collection of algebra or civics facts.

When algebra, science, or civics are treated as just a bunch of facts to be tested by paper and pencil tests, even the students who can write down these facts on a test often cannot use them to solve real problems. For example, even high school and college students with As in physics—students who can, for example, write down Newton's laws of motion—often cannot use those laws to solve problems in the world. In one study, the majority of the students in a college physics class could not say how many forces were acting on a coin when thrown in the air, even though the answer can be deduced from Newton's laws of motion and these students could write these laws down as equations on a test (Chi, Feltovich, & Glaser, 1981).

To a physicist, physics is not a body of facts and information. A real physicist knows many facts, but those facts are tools for solving real-world problems. Since the physicist repeatedly needs and uses such facts for problem solving, he or she remembers them well and with little effort.

So, too, is the case with civics. If we want people who can think about and solve problems in the world, then civics facts must be seen as and used as tools for solving problems—in this case problems of civic engagement, participation, and transformation. Similarly, students need to see facts in mathematics and science as tools to solve problems. Then they will have no trouble remembering them and knowing why they are significant because they will be using them repeatedly to achieve their own goals as problem solvers.

Here is one example from games, in this case a so-called serious game. David Williamson Shaffer, a learning scientist at the University of Wisconsin–Madison, makes what he calls "epistemic games" (Shaffer, 2005, 2007). In epistemic games, students take on the role of a professional and must learn, think, and problem solve like a professional using the tools of the profession, not just facts, but technological tools as well. The students learn to see the world as a particular sort of professional does. They take on the "knowledge frame," or what Shaffer calls the "epistemic frame" or "epistemic lens," of that profession. It is as if the students put on a pair of glasses that show them how the world looks and how you can transform it when you use the knowledge and tools a given profession has developed.

One of Shaffer's games involves middle- and high-school age students in the role of professional urban planners (Bagley & Shaffer, 2009). The game is not unlike the commercial game *SimCity*, where the player has to build and manage a city, except that it requires the students to use professional knowledge and digital tools in their urban planning. Furthermore, the game is both on a screen (where the students plan a part of the real town they live in and see the changes in a virtual 3D world) and in the real world where they visit real streets and, in the end, have to present and defend a real urban-planning report to real urban planners. The students use professional-like zoning tools and codes. They also, in the game, get feedback from a virtual mayor and other stakeholders (environmentalists, business people, citizens, and so forth) on how they feel about various proposed plans. A new parking lot may well please the business people but anger green-space advocates. Students have to balance a variety of social, environmental, political, and economic interests.

There are many facts to learn in becoming an urban planner, including many codes, or rules, for different zoning decisions. One traditional way to treat this information, a way we too often take in our schools, would be to have students read and memorize them. The students might or might not be able to apply these facts and codes if their school could spare the time, away from testing, to let them "play" as urban planners.

This is, of course, not how Shaffer proceeds. He demands that the students propose a solution to an actual problem in the town they are replanning. The students must be professionals and soon discover that the problem cannot be solved without using certain facts, codes, and technological tools. As the students use these facts, codes, and technological tools over and over again, the information becomes second nature to them. They learn all the facts and codes (or where to find them), the stuff schools too often consider "content," without really studying them in isolation. The information comes "free," so to speak, or without deliberate memorization or study, as it repeatedly gets recruited to solve problems the student passionately cares about. Shaffer's students do passionately care because they have to face a real urban planner and not appear foolish, and, in any case, they are replanning their own community, something they feel strongly about. They are doing what adults do, just as if they were adults, and this is highly motivating.

The fact that even commercial video games stress problem solving over content is not a problem. It is a virtue. It is, in fact, a perspective and attitude we badly need in schools if we want the kind of "deep learning" that we describe previously: solving complex problems, taking on new identities and worldviews, and using facts and tools in meaningful ways. At the same time, we must stop bemoaning the content of video games we do not like and instead start making games that involve problem solving we value, with problems that recruit facts as tools we value as well.

Thus, we do not want a game, or a textbook, for that matter, that treats civics, for example, as dead facts but rather treats civics facts as powerful tools for civic engagement, understanding, and transformation. For example, why is it important to know the three branches of the U.S. government in the United States (the executive, legislative, and judicial branches)? Just to pass a knowledge test? At best, this is school as *Trivial Pursuit*.

Imagine a role-playing activity, or a real problem in the real world, where you as a single individual must seek redress for a serious wrong. To seek help from the president (the executive branch), you must have a good deal of power or money behind you to get his or her attention. It will help to be a significant player in the president's party. To seek help from the U.S. Congress, you better be part of a significant interest group; your problem ought to be the problem of many people who are part of a well-defined group whose interests can be readily identified. The courts are the only place you can access and where you can stand as one individual and seek redress.

When you do seek redress in a court, whether in a game, a role-playing activity, or a real case, you will really care that the court is fair and neutral, free of political influence and not beholden to one political party. You will understand why it might not be a good idea to elect judges or have them run as Democrats or Republicans. You will understand why the Supreme Court is not elected and why judges serve there for a life time. You will understand how problematic it is that people, including the president and the U.S. Congress, seek to vet judges for either right wing or left wing political correctness. You may come up with different viewpoints on such issues, after reflection, but you will care, and the facts will now be lively tools for thinking through real problems and for real debates. Asking students to write down the three branches of government and giving them a good grade when they can will not suffice if we want real understanding based on mastering tools, and facts are one type of tool, for problem solving.

Cultivated Children

We conceded above that Sam's story was not universal. He was from a privileged family and both his parents were professionals. In a close ethnographic study of child rearing in different homes, Annette Lareau (2003), in her book *Unequal Childhoods*, has identified two different models of what it means to raise children. One model she calls the "cultivation model." This model is applied mostly, though not exclusively, to middle- and upper-middle-class parents. The other model she calls the "natural growth model." This model is applied mostly, though not exclusively, to those parents in the working class or lower classes.

Our research on games and learning in homes and communities has shown us much the same thing with games and other digital media. When parents hold the cultivation model of child rearing, they treat their child like a plant that must be constantly monitored and tended. They talk a good deal to their children, especially about topics that do not just involve the here and now. They use a good deal of "book language" and adult vocabulary around their children, especially in the areas where their children have become "little experts" (as Sam did on mythology), something these parents encourage.

Even though they are the ultimate authorities in their homes, these parents negotiate with their children so their children get lots of practice in developing arguments and explanations. They arrange, monitor, and facilitate a great number of activities for their children, such as museum trips, travel, camps, and lessons (e.g., music, ballet), and other special out-of-school activities. Through these acts, they heavily structure their children's free time (and, yes, sometimes overstress the children). They encourage their children to look adults in the eye and to present themselves to others as a confident and knowledgeable person, or at least a person with a right to an opinion. They encourage their children to develop mastery with digital tools, using things like games as a gateway, and help their child relate this mastery to literacy and knowledge development, much the same as we saw with Sam.

Cultivated children can be, in some cases, too empowered, perhaps even at times obnoxious. They can be overstressed and need more free time to just be children (or even childish). Regardless of what you think of such parents and their children, the evidence is overwhelming that the cultivation model is deeply connected to success in school and to aspects of success in society, at least at the level of income and higher-status jobs. This form of child rearing and its relationship to economic success may be a bad or good thing, but it will not go away unless we radically change how our society works. In the meantime, we better find ways to help children who do not get such cultivation at home.

When parents hold the natural growth model, they treat their children like a plant that, with rich enough soil and nutrients, can be left to develop naturally without a lot of extra tending. Such parents love their children and care for them deeply and well. But they do not feel the need to intervene constantly in their children's lives, from the earliest years on. Often they cannot intervene as much as more well-off parents because they are busy working and surviving. They talk less to their children and use less adult language with them. They tend to use more directives and commands with children and do not negotiate with them. They expect their children to be respectful and deferential to adults. They do not structure all their children's free time and expect them to learn to find things to do with their peers and by themselves. They do not attempt to direct their children's use of digital media (like games) toward other more school-based skills, engagement with computer software, or higher-order literacy skills.

Children raised with the natural growth model are often hard working, self-sustaining, and respectful. They are not always comfortable with putting themselves forward or presenting themselves as knowledgeable to adults, even when they are. They are not always comfortable with engaging in arguments, explanations, or sharing opinions with adults, especially those they do not know. They have not built up lots of language, experience, and knowledge connected to the myriad of activities children raised with the cultivation model have experienced.

Many children raised with the natural growth model have done just fine in school and have significant success in life. But if we look on a statistical level at group trends, they tend to do significantly less well in school and in society, at least in regard to income and positions of power and status, which, of course, are not the only, or even the most important, markers of success. We acknowledge, too, that the two models we have discussed are really poles of a continuum and there are parenting styles in between. There are, too, of course, parents who neglect their children and give them little or no care at all.

But despite all the reservations and concessions we can and must make, we have to face the fact that we have here an equity crisis. It is an equity crisis that is getting bigger and increasingly involves digital tools, including video games. Children from more privileged homes, raised with the cultivation model, are acquiring a myriad of skills, values, and attitudes that contribute to success in school and to certain sorts of success in society that should be open to everyone. Much of the mentoring and learning these children are gaining is at home with their parents and in interactions with many others, peers and adults, on the Internet, as well as face to face in the real world. Digital media, like games, along with many other digital tools, are making it easier to "cultivate" children and are allowing children to be better mentored, mentoring that now goes well beyond their parents.

Many skills these children acquire are not even part of the curricula in most of our schools, especially the schools many less-advantaged children attend. Digital media, when coupled with the cultivation model, can widen gaps in knowledge, literacy, and technological skills between rich and poor children. The existing gaps are bad enough. We need to ask ourselves how we can cultivate all our children, in and out of school, while at the same time widening our ideas about success beyond achievement in schools as it is currently defined and purely monetary success later. This will be an important theme we will return to in this book: what constitutes success in a globalized world where relatively few people can gain high incomes and powerful, high-status jobs?

Themes for This Book

The girls and women in this book, who are gaming, designing new aspects of game play and game content, and writing stories inspired by games, are part of a larger popular-culture trend today in which people produce, and not just consume, content. These girls and women do not just want to play games; they want to make things for and around the games. In this act, they are forming powerful learning communities from which we have much to learn.

So why have we started with Sam? What we have to say about girls and women in this book is not meant to imply that there is some inevitable divide between women and men's psychology or behavior in general, nor is it meant to imply that women and men always game in different ways. We will argue that Sam's story exemplifies changes in the nature of gaming that involve both men and women, but women are contributing to these changes in interesting and special ways. Women's contributions go largely unheralded, since the media pays much more attention to men as gamers than they do to women as gamers.

Sam's story shows that, today, gaming goes well beyond gaming. Gamers use games as a center for activities that go beyond just playing the game. They mod the games they play; they read and write about the games they play and about the themes these game involve; and they collaborate, often through the Internet, not just to play together, but to mod, to discuss the game, and to organize all sorts of interaction and learning around the game and gaming more generally. In these acts they pick up technical skills, design skills, and interactional and collaborative skills. These skills are important skills in the twenty-first century in which we live, a time replete with new scientific discoveries, rapidly changing digital technologies, complex systems and complex problems, and a myriad of conflicts.

These trends are playing out a bit differently, though with lots of similarities, between female and male gamers. One game that clearly illustrates some of these differences is *The Sims*. As we write this, *The Sims 3* has just appeared (and sold a million and half copies in two days—see Terdiman, 2009). *The Sims* is a game where the player builds a family, neighborhood, and community and guides the lives and destinies of virtual people. Will Wright and the team of designers who create and support *The Sims* have made a deliberate effort to allow and encourage players to build and design all sorts of things for the game, becoming producers and designers. While professionals who created *The Sims* maintain a popular community Web site for *Sims* players, the players themselves have organized many other collaborative-learning communities that, at their best, exemplify a model for the reform of twenty-first-century learning and education. In these communities, they organize and coordinate each other's play and creativity in ways that display a great deal of what has been called social and emotional intelligence (Goleman, 1995, 1997).

While we argue that girls and women playing *The Sims* and other games represent the cutting edge of these trends, we readily acknowledge that many male gamers act just like the women gamers we discuss, and alternatively, many women gamers do not. There is no "great divide," but just some suggestive trends. We were reminded of this point when we checked in with Sam and what he was playing as he turned fourteen. He had downloaded the *The Sims 3* for his iPhone, then bought the computer game and was happily playing both. We will argue in this book that many men are following in the direction where the women gamers are leading.

Chapter 3

The *Nickel and Dimed* Challenge

Designing New Forms of Socially Conscious Play

Movies, Books, and Video Games

Many people think that games are in competition with books. They believe that, as people play more games, they read less. In fact, gaming takes away the most time and attention from television, not reading (Zane, 2005). In many respects, young people today are reading and writing more than ever. They read and write about the games they play and the interests these games engender, as we saw with Sam and the game *Age of Mythology* in Chapter 2. A good deal of this reading and writing is technical, involving technical and technological aspects of gaming activities, such as ways to mod games. However, at the same time, fan fiction, devoted to almost any movie, book, or game one can imagine, is flourishing (Black, 2008; Hellekson & Busse, 2006; Jenkins, 2008).

Books and games, though, are very different media, much more different than many people who do not play games realize. In this chapter, we will eventually talk about a woman who was inspired by a book to make up a game and then challenge other people around the world to play it. But before we turn to this woman, we want to look at the differences between games and other media, especially books and movies. This will let us explore the significant differences between games, on the one hand, and books and movies, on the other. This will help give us a deeper understanding of what a game is and what potential games have for the sorts of deep thought that we know books and movies at their best can foster.

Think for a moment about cases where books and movies have been made into games and where games have been made into books and movies, something that is increasingly common today. Making a book into a movie, something we are all very familiar with, is not an easy task, as anyone who has seen a favorite book lose most of its substance in its translation to film knows, but it turns out to be much harder to translate a book or movie into a good video game. Though it is a rare action movie that is not made into a game these days, most video games based on movies garner poor reviews. One game site had this to say about games made from movies: "During the summer season of gaming many, many games based on big blockbuster movies find their way onto store shelves. Regrettably most of these games . . . end up on the wrong side of average. In an even worse turn of events, most of these substandard games then seem to find their way to the top of the charts, based solely on the selling power of the license, and we imagine then bring much sadness to those that chose to play them" (McDermott, 2007).

It is also common to make movies based on games, and these have fared even more poorly. Some have become famous for how dreadful they are. It has been said of the director Uwe Boll, who made a number of movies based on games, that his *House of the Dead* "defined horrible film for an entirely new generation" (Darnell, n.d.). The relatively recent movie based on the game *Max Payne* received this sort of review: "*Max Payne* is yet another film to join the ever-swelling ranks of game-to-movie adaptations that sound fairly promising in their conception, but end up being absolutely terrible" (Parfitt, 2008).

There are, of course, exceptions. The computer game *The Chronicles of Riddick: Escape from Butcher Bay* (and its platform version, *The Chronicles of Riddick: The Assault on Dark Athena*) received high praise from game reviewers and players. The game served as a prequel to the movie, *The Chronicles of Riddick*, which received generally poor ratings from film critics (an average 38/100 score on MetaCritic.com). In fact, some game reviewers explicitly noted the game to be an exception to the generally mediocre quality of games based on movie franchises (e.g., Kasavin, 2004). This example shows that what makes games good and what makes movies good are two very different things. There have been a few cases when games based on movies were quite well-received; one example is *Peter Jackson's King Kong: The Official Game of the Movie,* a first person-shooter, action-adventure game based on the 2005 film *King Kong.* It was a collaboration between the film's director Peter Jackson and famed video game designer Michel Ancel, which perhaps accounts for its successful execution.

Books and movies are highly valued as storytelling art forms. However, in regard to video games, the idea of story is controversial. Some gamers, game designers, and game scholars think that a good story is crucial to games because it motivates the player to understand and value the world he or she is in and the actions being taken, but others disagree. They say that a story is not important in games. What is important is game play, namely, the decisions the player makes, the strategies formed, the problems solved, and the actions taken. Stories are authored by other people; what is important in games is what the player does (for further discussion of these viewpoints, see Juul, 2005; Wardrip-Fruin & Harrigan, 2004).

Whatever side we choose on this controversy, it is clear that the role of story is different in a game than it is in a book or movie. When making up a story for a book or movie, the author can develop a beautifully structured narrative from beginning to end. The reader cannot change the story and must accept it as it comes from the author (although, of course, readers can interpret what they read in all sorts of ways). In games, players must act and make decisions. The author of a story in a game must allow for players to make choices and the story cannot completely determine these choices or foreclose all of them. Many games even have several different endings, contingent on decisions a player has made earlier. Players will not play a game just because it has a good story, and a good story cannot make up for boring or otherwise poor game play. Games are not only about the virtual characters in their stories but also about the player. The player is and has to be the "hero" (though the player is not always "good" and can sometimes choose to be good or evil).

Game designers are often storytellers, but they are primarily designers of the relationship between players and a game world by creating problems for the player to solve and by giving the player fun and interesting ways to solve them. If a good story contributes to this enterprise, it is more than welcome in the game. A poor narrative simply becomes irrelevant. When we translate a book into a movie or a movie into a book, we are translating from one story-driven form to another; we are only dealing with different formats for telling the same story. On the other hand, when we translate from a movie to a game or from a game to a movie, we are translating between a story-driven form (the movie) with an emphasis on the author or writer, and a problem-solving form (a game) with an emphasis on the player.

Translations of books into games are quite a bit rarer than translations of movies into books and also have not fared well at the critical level. Few people think *Harry Potter* video games are as good as the *Harry Potter* books or movies, though they have received some positive reviews (see Metacritic, n.d.). *Lord of the Rings* games have fared somewhat better among the critics (e.g., see Keefer, 2004), but these are really translations of the movies by Peter Jackson, not the books directly, and include lots of dramatic scenes from the movies.

What makes translations among books, movies, and games so challenging? Each medium—books, movies, and games—does certain things better than the other media. A good translator has to find a way to take a story in one medium and translate it to new one, knowing that some power of the original medium will be lost. The power of the new medium will have to make up for that fact and bring out new virtues in the story. Translation is a technological skill, since the translator has to think carefully about the power of different media, and an artistic skill, since the translator must think about how to retain and even enhance the core values and themes of the story.

Translating between books and games is, as we have said, perhaps the hardest translation of all. At some level, players' choices have to substitute for plot points in the book's story. If a game has a story at all, and today most do, even if that story is original and not a translation from a book or movie, the game designer still faces the basic translation problem. In our culture, books and movies are our premier storytelling forms, at least so far. Since games are about players, their actions, and their choices, game designers always face the problem of how to translate the narrative bits and plot points of a story into good game play, or at least to render them integrated with, and highly relevant to, that game play. This takes a particular form of intelligence.

One particularly striking and novel example of how a game designer faced this task is *Braid*. *Braid* is a fairly recent two-dimensional (2D) side-scrolling arcade-type game, a game that can be downloaded from the Xbox Live Marketplace, Microsoft's digital distribution service for the Xbox 360. The character you play in *Braid* has to move and jump through each level, just as in any side-scroller type of game (such as the *Super Mario Brothers* games). However, you do not just simply manipulate the character in *Braid*; you can manipulate time and change how all the other characters and elements in any level behave. Time, and how you can manipulate it, works differently in each level in the game.

As you move through the levels, you read parts of a story, an intriguing but maddeningly vague story about the character you play, his loss of a girl friend, and his search for a princess. In the game you solve sometimes very difficult problems to obtain puzzle pieces that, when put together, show images from the story. The story is fascinating but not easy to understand. The game play is totally engaging, but the problems to be solved can be very difficult, requiring a good deal of thought about how to manipulate both your character and time (that is, once you even understand how time operates on the particular game level).

What does the story have to do with the game play in this case? People who think that stories are irrelevant to games will say, "Nothing," and claim the story is irrelevant to the game play; it is just an add-on to intrigue the player. But many players of *Braid* struggle to relate the story to the game play. For example, when we played *Braid*, we finally came to feel both the story and the game play were about an obsessive compulsion to persist in the face of rejection and difficulty, but other gamers saw it differently. That is the problem with stories in games: if they are not just window dressing, the gamer has to be able to see how the story motivates the game play. In turn, that game play enacts core values or key themes of the story and different players will see this enactment differently.

We have made it clear, we hope, that stories and games exist in interesting tension with each other. This tension, while it creates difficulty, opens up new possibilities for art and expression. We now turn to a player of *The Sims* games who, inspired by a book, decides to make up a game. However, *The Sims* allows for quite different relationships between stories and game play than do many other games, precisely because it is, essentially, a simulation of real life.

Nickel and Dimed

A German woman who calls herself "Yamx" on the Internet read the best-selling nonfiction book *Nickel and Dimed: On (Not) Getting By in America* by Barbara Ehrenreich (2001). Reading the book has inspired her to make up a game—in a sense to engage in a "translation" of the book into game play in *The Sims*. At one level, this translation is a lot easier than most book-to-game translations. Ehrenreich's book is about real life, the lives of poor people, and *The Sims* is a simulation in which players can act out some things from the book. But there are problems here, nonetheless. First, acting out the actions and choices of poor people sure does not sound like fun game play or a fun game. Second, *The Sims*, though it is a life simulator, is a commercial game and not really a very good simulator for poverty.

Even more interesting is that Ehrenreich's book is itself, in a sense, a simulation. She does not tell poor people's stories directly. Rather, she goes out in the world and pretends to be poor, simulates what it is like to be poor, when she herself is not poor. Yamx wants to use a simulation (*The Sims*) to simulate a book that is itself a type of simulation. Yamx has the added burden that somehow all this has to end up being fun, because, in the end, it is going to be a game.

Let us first discuss Ehrenreich's book. One of Yamx's goals, though not her primary one, was to get people to read Ehrenreich's book. The book is a fascinating and compelling exploration of how difficult it is to be poor in America, difficult in the sense that it takes a lot of skill and intelligence to be poor and still survive. The book is controversial, however, because some readers found it patronizing that a well-off woman tried to live the life of a poor one. Other readers found Ehrenreich's attempts to live the life of a poor worker illuminating because she made it so clear how difficult a life this is and how resilient and courageous the women she worked with often were. The book was published in 2001, but sadly the intervening years have only seen the spread between the rich and the poor get worse, making the book even more relevant today.

In *Nickel and Dimed*, Ehrenreich goes undercover as a low-wage worker in Florida, Maine, and Minnesota. She wants to find out how unskilled workers make ends meet. She spends one month in each location, where she works full time and attempts to live only with the money earned in the low-wage jobs she can find.

In Florida, Ehrenreich gets a job as a waitress in a diner and rents a trailer nearby. Her income cannot support her as well as pay the rent, so she takes a second job working as a hotel maid. She cannot handle the physical demands of the two jobs and quits the maid job after only a day.

In Maine, Ehrenreich joins The Maids, a residential housekeeping service. Again she finds it will take two jobs to make ends meet and so takes a job as a dietary aide in a nursing home. She ends up working seven days a week. The housekeeping job pays poorly and is both physically demanding and demeaning. As a dietary aide, Ehrenreich sometimes ends up taking care of an entire Alzheimer's ward by herself and fears for the safety of her patients.

In Minnesota, Ehrenreich cannot find a place to rent because of Minneapolis's low apartment vacancy rate. She has to go to a motel and pay a good deal more than she would for an apartment. Many poor people end up in motels because they cannot afford the first and last month's rent, plus security deposit, that many apartments require. It can be expensive to be poor. Ehrenreich also takes a job at Wal-Mart. Her job is to put the clothes back on the racks after customers take them off. To get the job she has to take a mandatory personality test. In the test, she is asked if she agrees or disagrees with the statement "All rules must be followed to the letter at all times." Not wanting to look like she is trying to "suck up" to the company too much, she answers "strongly," rather than "totally." The personnel director takes her aside to discuss her "wrong" answer. The correct answer is "totally" (p. 124).

Ehrenreich's book details the mental and physical rigors of low-wage jobs and makes it clear that such jobs deserve a living wage. Her book makes it clear, too, that so-called low-skilled jobs require a great deal of skill in coping with time pressures, multiple jobs, physically demanding work, and demeaning customers and managers. Being poor is hard work, since managing life with little free time and even less money is a massively demanding task, and survival is always in the balance. In many respects, Ehrenreich herself just did not have the skills required.

Nickel and Dimed was made into a play (a type of translation that we did not talk about above), but the translation did not fare all that well. The *New York Times* had this to say about the play:

> In her book Ms. Ehrenreich, a writer and academic, says about America in the age of the bottom line and welfare reform, "The poor have disappeared from the culture at large, from its political rhetoric and intellectual endeavors."

The play, with its exhausted fast-food cooks, waitresses, house cleaners, big-box-store saleswomen, and their striving managers happy to withhold the last penny from those they supervise, should offer one way into that missing public dialogue. But despite several excellent performances by the cast and a live band that is fitfully effective, the production at the Bank Street Theater is less promising than it might be.

In basing the play on Ms. Ehrenreich's work, [the playwright] by necessity has condensed the material, but the individuals the reader meets in the book tend to emerge as familiar types on the stage: the wives bullied by husbands at home and male bosses at work; a wealthy woman who looks down on her help. (Stevens, 2006)

Yamx's Challenge

What would be a good translation of Ehrenreich's book into a game? We have to determine the point of Ehrenreich's book and how that point can be made in a medium that stresses players' actions and choices. One possibility is clear: we could have players in *The Sims* attempt to live out the sorts of lives Ehrenreich described in the book (e.g., the lives of poor and often-single mothers). Beyond the burden of making this fun, Yamx faces a problem that Ehrenreich herself faced.

Ehrenreich's book is not first and foremost about the poor women with whom she works. It is about her. She is not trying to give the reader the details of poor people's lives in full. She fears, perhaps, readers would not care, since many in the United States tend to ignore or dismiss poverty if we are not poor. Rather, by trying it herself, as a person who, like many of her readers, is not poor, and exposing her follies, failures, and struggles, she wants to give the reader the *feel* for what it is like to be poor. She wants to the reader to experience actual emotions about poverty.

Affluent readers of Ehrenreich's book probably cannot fail to imagine how poorly they would have fared had they tried to do what Ehrenreich did. This imaginative act, triggered by Ehrenreich's approach, is a way to get people to feel something when they are often numbed by tales of poverty. At the very least, the book prompts us to feel that being poor might just be too hard for us, though it is clear that Ehrenreich hopes readers will also feel a certain indignation at how poor people are treated.

Ehrenreich is engaging with "emotional intelligence" (Goleman, 1995); she is trying to get people to see other people's feelings and perspectives and to be able to put themselves into other people's shoes. She is trying, as well, to get readers to reflect on their own perspectives, the sources of these perspectives, and how they feel—and why—about things like poverty and low-wage jobs. We will see that emotional intelligence is at the heart of what Yamx does, as well.

From Germany, writing in English on the Community forum on Electronic Art's *The Sims 2* Web site (http://thesims2.ea.com), Yamx offers *Sims* players the following challenge.[1]

The Sims 2: Nickel and Dimed Challenge

The following is the post that introduces the challenge: "This challenge was inspired by, and is named for, the book *Nickel and Dimed* by Barbara Ehrenreich (which has nothing whatsoever to do with Sims, but is nevertheless highly recommended). The idea is to mimic, as closely as possible, the life of an unskilled single mother trying to make ends meet for herself and her kids."

> The Goal: Raising your kids successfully until they're old enough to take care of themselves. If you can get all children to adult age without anyone dying or being taken away by the social worker, you've made it. If you want a kid to go to college, they can, but ONLY if they manage to raise at least $2500 in scholarships, since you won't be able to help them. Also, they have to live with you until their late teens (after all, 13 year olds don't go to college). Once the age indicator reads "becomes an adult in 5 days" (or less), you can have them move to college, but not before that.

This may not at first seem like a translation of Ehrenreich's book, and yet, at a deep level, it is. Games require a challenging and engaging "game mechanic," that is a core set of actions players take (Salen & Zimmerman, 2003). Any translation of a book or movie has to find a good game mechanic that is engaging (makes for good game play) and represents in a deep way the theme or point of the book or the movie. What would be a more natural translation for Ehrenreich's book than to make the actions a poor person takes to survive into the core game mechanic?

The Sims lends itself to this sort of challenge, in some respects, very well. It is, after all, a life simulator. It is well suited to translating real-world actions into game choices and actions. It seems almost easy to translate the narrative bits of any real story into a *Sims*-type game by telling people to simulate that story in the game. When games can simulate real life closely, they are quite good at translating real life into game actions. Military games like *Medal of Honor: Allied Assault* and *Call of Duty* have no trouble bringing to life the drama and the horror of real-world battles like the invasion of Normandy during World War II. The player gets to do, in the safety of home, what real people did to their peril in the real world.

There are, however, two problems. First, acting out the actions of people in the real world does not necessarily make the game interesting. While shooting in a military game might be engaging, shopping for clothes when you have very little money is much less so. That certainly does not sound like a good game mechanic.

Second, while *The Sims* is a life simulator, it is not a very good simulator of poverty. It is a commercial game, and, because being poor is neither fun nor entertaining, *The Sims* tends to best represent virtual lives that are considerably more pleasurable. Many players put in an easy-to-use "cheat" that gives the player lots of money so they can avoid the tedium of having to work their way up in the world. The game certainly does not include much of the aggravations of real-world poverty; for example, you do not have to worry about how to pay the utility bill, and your children will not be accosted by street gangs on their way home from school. Because *The Sims* is not a great tool kit for building the sort of simulation Yamx's challenge demands, she creates many rules to render it a better tool. In fact, she builds a veritable rulebook. She uses *The Sims* not just as a game in itself, but as a tool with which she can make up elaborate rules of play. She is making up her own game. She is modding, not in the sense of changing software, but in the sense of changing the ways in which people interact with the game.

Yamx has players download a residential lot and house she created herself to greatly restrict their resources. She stipulates many conditions that limit what parts of the game can and cannot be used. Let us look at just a few of these rules, such as this one: "We're simulating the life of an unskilled worker, so obviously, there are severe restrictions on skill building. (This is not to mean that real life unskilled workers can't be creative or charismatic, but within the game, that would invariably lead to promotions, so it's banned.) These restrictions apply only to the Mom, not the kids." *The Sims* allows virtual characters to develop different "skills" in the world. The list of skills is quite varied, including cooking, mechanical, charisma, body/hardiness, logic, creativity, cleaning, walking, talking, potty training, painting, meditation, pool playing, yoga, parenting, fire prevention, anger management, lifelong happiness, physiology, and couple counseling. If, for example, a Sim gains mechanical skill, he or she can better repair broken things in the household. However, Yamx is aware that such skill building can lead too quickly and too automatically to the sort of life no unskilled worker gets without a good deal of struggle. So Yamx must structure how players of the challenge can engage with skill building in order to keep the simulation realistic—and we will see in a moment what realistic means here.

Here is another one of Yamx's rules: "No cheats (except move_objects to remove bugged items)—in particular no kaching, motherload, maxmotives, or anything that aids survival. (If you want to use a hack to make the phone ring only five times instead of twenty, that's fine with me.) Custom content is okay if it's things like recolors or hairstyles. No special objects that will make your sims life easier, like a bottomless fridge or items priced $0." Lots of *Sims* players use "cheats" to give themselves infinite money or other advantages so that they do not have to go through the drudgery of starting at the bottom and slowly working themselves up in the world. These cheats are not necessarily a bad thing. They allow players to play the game in the way they want (e.g., to concentrate on building houses and designing clothes without worrying about low-end jobs, tight schedules, or money running out). But, obviously, for Yamx's challenge, such cheats would ruin its realism, that is, its fidelity to the life of a poor and struggling single parent.

We can note something that is quite apparent in many of Yamx's rules: she is thoroughly knowledgeable about the technical details of *The Sims* as a game, as a simulation, and as a piece of software. She has considerable technical knowledge, but she is using it to build social engagement with both the game and issues of poverty.

Here are a few more of the many rules Yamx constructed to make her challenge realistic:

Life in poor neighborhoods is often unsafe—you cannot have a burglar or smoke alarm, and no sprinkler.

If you're given the genie lamp, you have to keep it, unused, in some corner of your lot. No selling it, and no wishes.

No quitting without saving after bad events.

The last rule above is a tough one for gamers. Gamers, young and old, often just turn off a game without saving what they have done when things have gone wrong and they do not want to live with the consequences. Obviously poor people cannot do this in real life, and players taking Yamx's challenge cannot do it either.

Even though Yamx is designing a rule kit, she is not a "boss." Players regularly write to her in the thread and dispute her rules. In turn, Yamx defends her rules or changes them. For instance, some players write back to Yamx telling her that her restrictions are not realistic in terms of how poor people actually live in the real world, as in the post below: "Hi, I am about to start this challenge. I just have one thing to say before I do. Being a poor person, I am very well qualified to say that no smoke detector is completely unreasonable. They are required by law in all low-income housing and given away freely by fire depts. to poor residents for the asking, at least in the USA" (mrskathycohen, 2008). Yamx's response to this post suggests what counts as "realism" in her challenge. We begin to get at the heart of what she is doing. Her response, in this and other such cases, shows us that she is focused not on a set of rules that are fully realistic in terms of the real world, since this cannot be achieved in *The Sims* or, for that matter, in any simulation. Any simulation is a simplification of reality made to help us understand issues that are often too complex to be dealt with in all their complexity all at once. Yamx is focused on modifying how people play the game so that their game play captures the degree of challenge in a poor person's life and "the feel" of being a poor single parent:

Oh, I realize that in real life, there are fire alarms in poor neighborhoods. Just as in real life, poor people can be terribly creative, and charismatic, and great cooks, and . . .

The point is that within the game, many real-world difficulties simply don't happen. There is no rent—everyone owns their house, no medical bills, no vandalism (except maybe someone kicking over your trash can), you can get away with having only one outfit, which never needs to be washed or mended, your children automatically get a full set of new clothes when they outgrow the old . . .

In a nutshell, the life of a poor sim is still MUCH easier than the life of a poor person (as you know). So therefore some of the rules are there simply to make sim life harder, and thereby bring it a little closer to real life.

Also, the fire alarms in the game work unreasonably well. Where on earth do you find one in real life that has the fire brigade on your doorstep in ten seconds?

There are many queries and responses like this on the thread. Yamx makes it clear that she knows "the challenge can only mimic the real thing to a small extent" and that she is trying to generate a feeling of "barely scraping by" and "of having very few options and just having to put up with a lot of stuff that was so apparent in the book." Yamx is using *The Sims* as a tool kit to design a simulation that creates the "feel" of being a poor single parent, not to engage in a realistic depiction like a documentary or realistic novel. Her simulation is a game with consequential decisions and problems to be solved by the players and a clear win state delineated in her rules.

At the heart of game design is the skill of finding a core metaphor or theme that can be translated not just into a story, but into good game play involving actions, decisions, and problem solving (Fullerton, 2008; Schell, 2008). Yamx has done just this. She has translated Ehrenreich's book into rules for game play that gives players the "challenge" and the "feel" of being a poor, single parent. The point is not total fidelity to life, but fidelity to feeling. In the act, she makes *The Sims* a good game and simulation for her purposes when it is not, by itself, good for those purposes. She has repurposed *The Sims* as a tool for the understanding of real life at the level of game play and not just story.

Modding

Yamx has modded *The Sims* to create a new game. She also has translated a nonfiction book into a game at the level of feeling and theme. Her modding and translation make her a game designer who uses *The Sims* as her engine (i.e., the software used to make a game).

However, Yamx's modding and translation are both, in certain respects, unique. She is not modding in the "traditional" way by creating new game mechanics or new game content through (re)programming software. This is the sort of modding, often done by males, to which people have paid the most attention, in part because they believe it can lead to high-tech skills. In this sort of modding, a young person uses the same sorts of software by which a game was made to modify the game. For example, a player may make a new skate park in a *Tony Hawk* game or even change a game about fighting aliens into a game about World War II. It takes technical skills and patience to learn to do this.

Yamx is not changing *The Sims* software, though she has to be deeply familiar with its properties. She is modding by creating new rules of play, or new ways of playing, for a wide community of people. She is socially organizing this community to play in a certain way, to think and reflect in certain ways, and to relate to each other in certain ways as they take her challenge, negotiate over it, and comment on it.

Like the celebrated game *Counter-Strike*, created by a group of young men who turned the game *Half-Life*, a science fiction first-person shooter video game, into a new game about modern warfare, Yamx's game is a mod. But it is a very different sort of mod, requiring a different sort of modding than *Counter-Strike*. We tend to focus on the *Counter-Strike* model and not the model that Yamx's challenge represents. We know that people who can mod in the *Counter-Strike* sense are digitally savvy, and educators would like to spread such skills. However, Yamx is digitally savvy in a different way, not just because she has exemplified high-tech skills, but because she has orchestrated a new combination of game play and social interactions around that game play.

Yamx is modding in a way that centrally involves both what has been called "emotional intelligence" (Goleman, 1995) and social intelligence (Goleman, 1997). For us, emotional intelligence (which we argue can be learned and developed) is the ability to understand and manage one's own emotions, decisions, and perspectives, as well as to understand and helpfully change or collaborate with other people's emotions, decisions, and perspectives to make the world a better place. Social intelligence is the ability to understand and manage social relationships and to collaborate with others. Emotional intelligence and social intelligence go beyond understanding facts and "reality" to understanding how to leverage facts and reality to engage with others in effective (and affective) actions and interactions. "Cognitive intelligence," the mastery of facts and information, is sterile and sometimes even dangerous without the guidance of emotional and social intelligence (especially in our highly complex and high-risk globally interconnected world). We can see Yamx's emotional and social intelligence, and her ways of fostering emotional and social intelligence in others, if we look at some of her many roles beyond designing the challenge.

Yamx's Roles

In addition to being a modder and translator, Yamx is a sort of "dungeon master." In live role-playing games (e.g., *Dungeons and Dragons*), where people enact fictional characters in face-to-face play, the dungeon master creates and elaborates story elements with and for the players. The dungeon master also adjudicates rules of play if and when any debates come up among the players. Dungeon masters monitor and enhance play moment by moment (see Fine, 1983, for an excellent analysis of face-to-face fantasy role-playing groups).

Players who attempt Yamx's challenge consult with her about the interpretation and limits of the rules, as well as the possible modifications they can make. She not only makes up rules of play but also continuously monitors how people are playing as they consult with her as "the expert." For example, consider Yamx's responses below. The first is a response to a player who put her (Sim) children into a private school (and propositioned the headmaster):

> Sorry, no private school. In real life, wohooing with the headmaster alone wouldn't get you in—he'd have to explain to the school board why your kids are not paid for, and what's he going to say—"Oh, I wohooed with their mom?" LOL.
>
> Even if there are isolated cases where something like this might happen, the simple fact is that the overwhelming majority of kids from poor families simply have no shot of getting into private school. (Hey, I thought I was being generous by allowing college! ;D)

The second is a response to a player who fears she has broken some of the rules:

> Aw, well, as long as you had fun with it, don't worry about the score/having accidentally broken some rules. It's not like it a competition, so you're not cheating it's all about having fun with a new way to play the game. :)
>
> And as you said, you can always play again, without using aspiration rewards, and with the "quitting when elder" thing—that way, you can claim the first time was just a test run. ;) Five kids all by herself during the depression? Wow. That's impressive. Maybe I'll try that next time.

It is apparent from the second quote that in addition to her role as a sort of dungeon master, Yamx is also a kind of coach or mentor, encouraging and rewarding the players with praise and helping them to see their play in a certain way: "It is all about having fun with a new way to play the game."

In addition to modder, dungeon master, and mentor, Yamx has yet another important role. She is also a kind of teacher. Along with her challenge, she gives the players an "assignment": "It'd be really great if you'd make a story to let us all see how it goes for your sims (lots of pictures are good), or at least post here and tell us about it. Here's mine: http://tinyurl .com/26fr58."

The Sims allows players to make an "album" that works something like a storyboard for a movie. The player can annotate pictures, creating an illustrated story (see Figure 3.1). Yamx not only gives this assignment but also offers her own story as an example. Throughout the thread, Yamx reads and encouragingly comments on people's stories (and they avidly seek her feedback). For players who do not know how to create stories and upload them, she provides a link to a tutorial and offers them guidance,

Figure 3.1 A player-created image from a *Nickel and Dimed* story, designed by Rachael-Ann Lyra Bellamy

Originally posted at *The Sims 2* Story Exchange (http://thesims2.ea.com/exchange/story_detail.php?asset_id=221257 &asset_type=story&user_id=2549855).

encouragement, and support. She is a teacher in the sense not of telling people what to do, but in the sense of encouraging and resourcing their own creativity and productivity. This is very much a role we want for twenty-first-century teachers in our schools (for an interesting point of comparison, see The Partnership for 21st Century Skills 2007 report, *Teaching and Learning for the 21st Century*).

Yamx is orchestrating particular types of social interaction in her roles as dungeon master, mentor, and teacher. One aspect of this social interaction that is particularly interesting is the way Yamx's mod and *The Sims* as a game interact to create a fascinating play of identities. The players, as they write their stories and in their reports to Yamx on the forum thread, engage in a type of writing that integrally mixes (a) storytelling, (b) details of their game play as determined by Yamx's challenge, (c) how aspects of *The Sims* as a simulation and game intersect with that game play, and (d) their real-life histories. In this complex writing, we see a play among various identities: the virtual characters' identities (the identities of the virtual characters in their *Sims* play sessions, characters with which the players very much identify); the players' real world identities; the players' gamer identities as gamers and *Sims* gamers; and the players' out-of-game, virtual, social identity on the thread discussing *The Sims* and Yamx's challenge with other people from across the world.

Consider, for example, the following from a player's story: "Poor Deb. She gets a job and the Nanny doesn't show up in time, so she has to quit and wait till the next day to look again. The funds are really getting low at this point, and we can't sell unused items or dig for treasure. I'm starting to freak out at this point, and so is Debsa."(Anonymous, n.d.)

Waiting until the next day to get a new job is one of Yamx's rules, as is not being allowed to sell unused items or dig for treasure. The "we" in "we can't sell unused items or dig for treasure" betokens the player's identity as a gamer playing Yamx's game. The clear identification of the player with her virtual character "Debsa"—for example, "poor Deb" or "I'm starting to freak out at this point, and so is Debsa"—melds her gamer identity (she is "freaking out" as a gamer; she may blow the challenge) and her virtual character's identity (her character is "freaking out" because her virtual life might end badly).

Or consider this from the same player:

After one day of working with a nanny and a toddler, she quit her job until the wee one was in school. Most of her money was going to child care, and she was exhausted. When the older one became a teenager it was much easier, and once there were two teens it almost felt like cheating, it was so much better. Luckily, Sim teens don't have gangster friends or drug problems.

I ended up getting a bunch of "pity promotions"—she ended up at third level because of chance cards. At that point I decided to make her quit due to "depression." (And she comes by that depression honestly. She's a family Sim with no real time to date and fall in love, and her aspiration of marrying off six kids just ain't happening in this lifetime. She spent most of her time in the red, and cried and worried a lot.) Then, of course, she had to start over.

Note how the player talks simultaneously about what she did as a gamer (e.g., "I decided to make her quit"); what the game, *The Sims*, does as a piece of software (e.g., chance cards and aspirations are part of the game mechanic of *The Sims*); the context of Yamx's challenge (e.g., "it almost felt like cheating"); and the story of what happened to her virtual characters (e.g., "After one day of working with a nanny and a toddler, she quit her job until the wee one was in school"). All these forms of writing are integrated seamlessly.

Consider this contribution to the thread:

Well I am finished with it.

Lily had to quit her job again and she was so depressed. I felt bad for her. Then the son got fired for making the wrong choice on one of those card things. Then he got depressed. I was not happy with him. His grades fell to A−. I had to have him walk to town and party for a while to bring him out of his depression. Then he did his home work and went to bed really late. So the school bus came and he didn't want to go to school but I sent Lily in to talk to him. He ran and got on the bus just in time. And when he came home his grades went to A.

He was running out of days before adulthood and Lily gave him a talking to about how much college would help and how much she wanted him to go. He went to school and came home with an A+ and got a scholarship. But not enough and only one day left. He was so worried. He loved his mom and didn't want to disappoint her. So that night he ran away and went to college. He plans to work at the coffee house to earn his way and still try to make the deans list so he can stay another year. I wish him luck.

Here again, we see a complete melding of talk about the story in the virtual world and the decisions and actions the gamer took playing both *The Sims* and Yamx's game. We see again, in phrases like "I wish him luck," the melding of a player's real-world identity and a virtual character identity.

The gamer talks about the virtual character as if he were real and a person in her real life and even attributes motives and desires to him: "He loved his mom and didn't want to disappoint her."

Finally, consider the two posts below where players clearly bridge the game (both *The Sims* and Yanx's challenge) with their real lives. The first: "This is pretty true to life! And I love it. When my son is older, I'll show him this as a story, he's only 11 months right now . . . So yeah! But I love that there are people out there who truly understand and have empathy for single parents, moms or dads." And another:

> So, even though this is . . . wow . . . so much like history repeating itself (I had a child and a toddler when I became a single mom; then I had a baby and had to go back to work in 7 weeks—and had to pay $125 a week for a sitter back when I was making only 10 bucks an hour. It was . . . interesting. We made it though. My daughter's going for her master's, my son is a musician writing his own stuff and my youngest son just got high marks and is about to enter college. So, the single mom gig is tough—but it's doable."
>
> This might be a nice way to tell my kids a bit about their history in a Sim-ish kind of way. :) Very very well done challenge Yamx. Kudos!

Note here how the trafficking between the virtual world and the real world goes in both directions, and neither side is privileged. Yamx's game reflects real life and, in turn, the stories and play sessions her game has generated can be used to inform people in real life. The players' own play sessions and virtual stories are treated as legitimate representations of reality, right along with novels, movies, and real-life storytelling, each with their own "truth" to tell.

Emotional Intelligence Modding

Yamx has done something that many professional game designers would love to do. She has made engaging game play from aspects of life that are often difficult and boring. At the same time, she has engaged players, as individuals and as a group, with reflection on the real world, on ways of representing that world, and on complex issues like poverty. This is designing with and for the emotional at its best (remember that emotional intelligence is closely connected with social intelligence).

We have done a good deal, and rightly so, to celebrate good game designers and good modders as tech-savvy creators. Yamx is being creative and tech savvy at a different level. We may miss her accomplishment because the male paradigm of game design and modding can lead us to see what Yamx is doing as "soft." It is "soft" in the sense that it uses so-called soft skills in addition to good technical skills "Soft skills" is just another term for a person's social and emotional intelligence. We might even call what Yamx has done "soft modding" if the word "soft" did not also have such negative connotations. Thus, we will call it "emotional intelligence modding."

Yamx is a game designer, modder, mentor, teacher, and leader, blending aspects of *The Sims*, game play, and emotional and social interaction all together in a coherent way. She is encouraging players' creativity and guiding them toward resources that can help them be productive. This is very much what we want from tomorrow's teachers and leaders.

There has been a lot of talk and effort given to the design of so-called serious games, or games for non-entertainment purposes (Michael & Chen, 2005). One problem with such games has been the difficulty in truly marrying good game play and content (e.g., algebra or health). Too often in such games, the game play has little really to do with the content. Another problem is that such games have not always been fun or truly engaging.

Yamx has much to teach us here. She has truly married her game play and content. Being a single parent becomes simultaneously content and game play at the level of gaining "a feel" for what it means to be poor in America. She has translated a serious book not only into compelling and engaging game play but also into engaging collaborative writing and social interaction that is integrally linked to that game play. She has shown that, if we want to make "serious games," the we ought to take into account not only the technical aspects of the game itself but also the social interactions around the game and identities it recruits in order to create games that turn out to be serious and entertaining at the same time.

Let us close with a quote from Yamx. One player says to her that it must be weird to see her creation all over the place. She responds, "It *is* weird—but a good weird, lol—here I am sitting in my little apartment with my cat, clicking something together on my computer, and suddenly dozens of people, some of them thousands of miles away, are using it, and enjoying it, and talking about it. :) It's really cool indeed."

Chapter 4

A Young Girl Becomes a Designer and Goes Global

Succeeding at Twenty-First-Century Skills but Not at School

Tech-Savvy Girls and *The Sims*

In this chapter we will meet a young girl named Jade (we have changed her name for privacy). Jade was part of an out-of-school Tech Savvy Girls Club (TSG; Hayes & King, 2009). The club was started with MacArthur Foundation funding, and had the goal of getting and keeping middle school girls interested in learning about digital technologies, particularly technologies that they might not otherwise find appealing. Lots of research has shown that, for boys, video games can serve as a gateway to technical skills. It is not just that boys play video games. Many of them eventually develop an orientation to games and other digital technologies that involve tinkering, modifying, "getting under the hood," and producing digital artifacts—games, art, and software programs—themselves. From game-related activities, these boys acquire wider interests in computers and other digital tools, and some go on to major in technical fields in college (Margolis & Fisher, 2002; Tillberg & Cohoon, 2005).

It is not that modern girls do not play video games. There is plenty of evidence that young girls like and play such games. But many of them lose this interest around middle school (Barker & Aspray, 2006; Gorriz & Medina, 2000). This is the same time that many girls give up or hide their interest in things like science and mathematics (Barker, Snow, Garvin-Doxas, & Weston, 2006; Goode, Estrella, & Margolis, 2006; National Science Foundation, 2007). As boy-girl relationship dramas become the center of middle school social groups, it is still, for too many girls, "uncool" to be a "techie" and smart. It is hard to believe this still happens in the twenty-first century and at a time when girls are succeeding more than ever before in school and beyond, but it does.

TSG was formed in response to a national concern. Women not only were underrepresented in fields like mathematics, engineering, and computer science, but also, in computer science, their numbers were declining (National Center for Women and Information Technology, 2009). Colleges across the country were worried about the lack of women in their technical majors and were trying a myriad of different approaches to solve the problem. Few seemed to yield much lasting success.

TSG was intended to create a place where it would be "cool" for girls to play games and get interested and passionate about technology. However, unlike many other such projects across the country, and there are a great many, that start with learning how to program (Hayes & Games, 2008), TSG started with a digital tool very much associated with girls and women: *The Sims*. *The Sims* is about building communities, social interactions, relationships, and virtual lives inside a simulation.

The choice of *The Sims* is controversial. Girls already seem to be adept with digital, social-networking tools (like MySpace and Flickr; Lenhardt & Madden, 2005). But, they do not seem to be as comfortable or proficient with the high-tech computer skills that lead to careers in science and technology. So why give the girls a game that focuses on the very social relationships they already are drawn to in such a big way? Shouldn't they be programming, hacking, and modding? Worse yet, *The Sims* involves players in designing clothes and houses and in having children and building families. Isn't this just reinforcing stereotypes about girls and women?

The Sims, when players become passionate about it, is not just about families, social interaction, and designing things. It is also about twenty-first-century skills. *The Sims* can create a distinctive approach to technology-related learning, one that melds technical skills with emotional intelligence. We argue that this approach, far from being just a "woman's thing," is essential for everyone in our high-risk, high-tech, complex, world. It is easy to miss the importance of this point if we see *The Sims* just as a "girls' game" and dismiss it.

Emotional intelligence is the ability to manage emotions, relationships, perspectives, and social interactions in ways that make for effective and ethical action in the world (Goleman, 1995). We live in a world replete with risk and complexity from major global economic, environmental, and cultural problems. In such a world, technology and science, as well as government policy, badly need to be under the guidance of emotional intelligence if we are to work together and not against each other. We do not believe emotional intelligence, any more than cognitive intelligence, is purely innate. It can be learned and enhanced.

Jade

One afternoon, as the women who coordinated TSG watched Jade's movie, one thing was clear: she was tech savvy. Jade's movie featured Sims wearing vintage fashions that Jade had designed herself. She had taken pictures of real clothes from the Internet, used Adobe Photoshop to modify them to her own taste, and transferred them to her game. She could then dress her Sims in these custom created clothes for the movie. She had not, of course, used a video camera or film to film her movie. Rather, her movie was a "machinima," that is, a movie made with a game engine. Creating this machinima involved using a built-in movie-making tool in *The Sims* game as well as other video-editing software tools and various cheats to modify the movie "set" and Sim actions (for examples of machinima made with *The Sims*, see Electronic Arts, 2009, as well as many that can be found on YouTube).

The skill required to make a machinima with *The Sims* movie-making tool is both technical and artistic. Jade did a type of modding to create her machinima, the very sort of practice that many consider a key gateway for the development of high-tech skills among boys.

In her time in the club, Jade developed skills that were now beginning to push her beyond many of the other girls in the club. However, at the beginning it was by no means apparent Jade would be a star. She did not have a family that actively encouraged her to develop technical skills or that emphasized education. In fact, she was not closely affiliated with school and did rather poorly academically.

In addition, when Jade started TSG, she was the least enthusiastic of the girls about *The Sims*. Like many girls, she had played the game before in an earlier version. The game no longer challenged or excited her. But in the club she would learn something new about being a gamer. She would learn to become proactive in her approach to *The Sims* and then in her approach to other digital tools. She would learn to be a producer and not just a consumer. Ultimately this learning would change her identity. She would come to see herself differently.

Jade is from a two-parent, white family in a blue-collar rural community. She is the youngest of three daughters and has a close relationship to her sisters. Once, after joining TSG, she was with her father visiting her oldest sister at work in a local Pizza Hut restaurant. As she watched her sister serve food, she told her father that she would never work in fast food restaurants. She felt that with the technical skills she was getting through TSG, and the skills she now hoped to gain in computer classes at school, she would never have to. Her sister herself had sworn she would never work in fast food, only to be forced into food service years later due to her limited schooling. For Jade, even with her new skills and ambitions, the question still remains whether skills gained out of school can sufficiently change her affiliation with school or even bypass school altogether as a route to success in society. We will return to this issue eventually.

Jade Before the Club

Before joining TSG, Jade was an active computer user. She sent e-mail, surfed the Web, and was heavily involved with MySpace. She played online games, including *Heroes Online* (http://hero.netgame.com/) and *Neopets* (www.neopets.com/). She had also played the original version of *The Sims*. She played the games with her sister, her brother-in-law, and a male cousin about her age who had played *The Sims* for some time and who showed Jade how to use "cheats" in the game (e.g., how to give your Sims lots of money so you can concentrate on building and buying, and not the drudgery of working your Sims up in life from menial circumstances).

Jade's family had high-speed Internet access and placed few restrictions on her Internet usage. However, the computer was placed in the dining room of the house, which was in easy view of her parents. Jade used the Internet daily, predominately for MySpace, though she also visited fashion sites, gaming sites, and blogs. She wrote to her MySpace blog twice a week and "hung out" on her MySpace page every day. Her MySpace site was the hub for her social-networking activities.

When Jade entered TSG and was asked if she considered herself to be a "gamer," she did not consider herself to be one. She said that a gamer was someone like her brother-in-law "who plays 24/7." (All quotes are from interviews with Jade while she was participating in TSG from 2007 to 2008.) However, the TSG Club eventually began to elicit some of this late-night gaming behavior out of Jade herself. When she got interested in creating clothing in *The Sims*, she eventually spent entire weekends on the computer and often stayed up most of the night.

The science educator Andy diSessa, a physicist by training, has argued that what prepared him for doing well at physics in school was the tinkering with mechanical devices he did with his dad in the garage. This tinkering did not teach him physics. Rather, it made him confident in his ability to handle technical and technological learning when he confronted physics in school (diSessa, 2000).

Contrasted with diSessa's experience, the ways in which girls like Jade play *The Sims*, blog, and interact socially on sites like MySpace are not seen as foundations for later important technological learning in the way tinkering in the garage or boys' game-related hacking practices often are. But we will see that they were just such a foundation for Jade. Social networking can motivate girls like Jade to learn to use technology to transform their social interactions and their own identities.

Getting Into *The Sims*

Although Jade had played games at home, including *The Sims*, when she entered the TSG Club, she could not get engaged with game play in the newer edition of *The Sims*. She no longer enjoyed just playing the game as it was. But then something important happened. The club's leaders, Betty and Beth, told the girls that they could, if they put in the effort to learn, actually design clothes in *The Sims* and make clothes that looked any way they wanted. This set off a spark in Jade. That spark would turn into a fire and change the trajectory of her life. Jade saw a new niche for herself in the game and the club as someone who produced and did not just consume. She was about to discover her passion.

Passions do not just come out of the blue. Jade had a long-standing interest in art. She once hoped to be an interior designer when she graduated from college. She had used computer drawing tools in a project for her 4-H club and had, in fact, earned a blue ribbon. When she was younger she had used Paint Shop, a computer art tool that came with her computer. And, of course, fashion is not an uncommon interest for middle school girls. However, in real life, Jade was not a girl who seemed fashion conscious. She frequently wore a baggy sweatshirt and jeans instead of more trendy clothes, which she was, however, often wearing under her sweatshirt. One of the other girls in the club often chastised Jade for not taking off her sweatshirt and simply wearing the "cute" shirt underneath.

To the club coordinators, Jade commented on her feelings about wearing plus-sized clothing. She regarded the skinny girls as the "pretty ones." When she eventually became a designer in *The Sims*, she started by designing for slim, model-sized (virtual) girls. But eventually she found a player-created download that allowed her to design clothes for plus-sized Sims. Since she was already adept at design at that point, this allowed her to create a whole new line of clothing tailored to sizes that are not typically reflected in *The Sims* games.

We can look into Jade's background and see an interest in art, some early experience with computers and social networking tools, and an interest and concern about fashion and bodies. None of these had heretofore led her to a passion, except perhaps for social networking on MySpace. We do not know how these factors contributed to the passion she eventually developed for design and, later, for computers as a source of power. The fact is that we know far too little about how passions develop. Jade was, perhaps, prepared for her passion, but she still needed to find it as well as to be mentored and resourced to develop it.

Passions matter in a big way in today's world. In our globally interconnected world, jobs that require only standard skills will often be done in lower-cost locations like China and India. It does not matter whether these jobs are low status (such as working in a call center) or high status (such as computer engineering or radiology), as long as they are not dependent on face-to-face contact. Thanks to modern technology, workers in such jobs can be trained and do their work almost anywhere in the world. In a developed country like the United States, rewarding careers will require the ability to innovate and not just apply standard skills.

Being able to innovate requires mastery. Mastery requires thousands of hours of practice (Gladwell, 2008). Mastery and thousands of hours of practice require passion. Otherwise, people give up. In fact, success today requires a disposition that has been called "grit" (Duckworth, Peterson, Matthews, & Kelly, 2007). "Grit" is the combination of persistence plus passion, and it is the passion that causes the persistence. Duckworth et al. use "grit" to mean perseverance and passion for long-term goals. We want to mean something just a little bit different. By "grit" we mean passion for and perseverance or persistence in problem solving in a given area or domain. We will use the word "persistence" instead of "perseverance" to distinguish our use of the term from Duckworth et al.'s because we want to stress that problem solvers with grit continue on not (just) because of external goals but because of their passion for the area or domain in which the problems reside.

Becoming a Designer

Betty and Beth told the girls, as we mentioned previously, that they could design clothes for *The Sims*. If we make traditional assumptions about teaching, assumptions we have picked up from school, it will seem unfortunate that Betty and Beth did not know how to design clothes themselves. We regularly expect that teachers know everything in advance, and then "instruct" learners by telling them what they need to know.

However, it was fortunate for Jade that Betty and Beth did not know how to design for *The Sims*. They could not teach it all to Jade and take away her ability to direct her own learning and passion. Rather than instructors (tellers), they had to become mentors and resourcers for learners who needed to operate under their own direction.

We argue that this is what real teaching is or should be. People do not learn (at least not in any way that changes them over the long term) by being told everything. People learn because good mentors give them some fuel for their engines and then they drive themselves, eventually learning to fuel themselves as well.

Only Jade initially took up the invitation. When Betty and Beth told the girls they could become designers, Jade was excited. The others girls were interested, but not enough to become passionate about the idea. Eventually it would be Jade and her success, not Betty and Beth, who motivated the other girls.

Jade no longer wanted just to buy clothes in *The Sims*. She no longer wanted to be just a consumer; she wanted to make things. She wanted to be able to take pictures of real clothes, digitally transform and redesign them to her own desires, and put them on her Sims. She wanted to become a designer. This is a crucial point about real education: it is not first and foremost about knowledge, but about becoming something. What Jade wanted to do could be done with software like Adobe Photoshop. But the price was learning a very challenging piece of software, and learning it well.

The price was higher than Jade could have realized in the beginning. When you start to design clothes for *The Sims*, you eventually discover, as did Jade, that there is a worldwide community of *Sims* designers, who design clothes, houses, furniture, environments, and even graphic novels and movies. This community exists all over the Internet. If you really want to be part of this community, and want respect and status in it, you have to go well beyond being "adequate" as a designer. You have to be good. The community has high standards and does not accept mediocre work.

To help mentor Jade and the other girls as beginning designers, Betty and Beth decided to work through an introductory tutorial on the Web that describes how to use Body Shop (a piece of software that comes with *The Sims*) and Photoshop together to recolor clothes that already exist in *The Sims*. The tutorial was on a fan site, *Mod the Sims* (see Faylen, 2006). Betty and Beth worked through the tutorial together. They did not find it easy, and had to help each other through it. At the time, they were using *The Sims 2*. In the newest version, *The Sims 3*, a built-in tool allows players to recolor clothes without this kind of effort. However, fans continue to mod many other aspects of *The Sims 3*.

Then Beth showed the tutorial to Jade, since she had shown the most interest in designing, and helped her work through it. After that, Jade took off on her own. She went home, did other tutorials, and spent hours practicing. She had never devoted this much time to *The Sims* or to gaming before. As she learned to create new things, she showed the other girls in the TSG Club how to do what she had learned.

Even a seemingly simple task like recoloring clothes (in *The Sims 2*) involved many steps. First, the player used Body Shop to select a preexisting Sim clothing item for modification. A particularly important part of mastering the use of this tool in connection with Photoshop was learning how to save files in the correct format and location so they will be recognized by both software programs.

When you open a clothing object file in Photoshop, whether an item from *The Sims* or a photo of a real piece of clothing, it does not look anything like the original item. For the purposes of editing, the item is flattened and spread out (i.e., in two dimensions [2D]), something like how a piece of fabric would look when you cut a dress pattern from it. In the initial tutorial Beth introduced to Jade, the entire dress was simply changed to a different color. Using subsequent tutorials, and through trial and error, Jade learned how to go much further. She learned how to recolor selective parts of a clothing item, how to change the texture of clothing by importing picture files she found on the Web, and how to change the shape and style of clothing by editing several layers of the clothing "mesh." (Mesh files determine the shape and other properties of an object.)

Jade learned how to use photos of clothing she found online to create custom clothing (see Figure 4.1 for an example). This is done by editing the photo, overlaying it on a plain *Sims* clothing item as a base, and then cutting away or coloring the base to make the clothing item "fit." She learned how to add logos to t-shirts in a similar way. Initially, these were time-consuming tasks that required much attention to detail, but eventually Jade became quite adept at these modifications. She also became a perfectionist, experimenting with different versions until she came up with something that met her own standards and eventually the standards of the wider *Sims* design community.

Jade did most of this learning on her own, with periodic support from Betty and Beth. She was spending a lot of time on the computer. This was, at first, a source of conflict with her mother. Jade's mother thought Jade should not be spending so much time "playing" on the computer. Like many parents, Jade's mother did not know the difference between being a consumer and being a producer in regard to games. She did not see that her daughter was building artistic and technical skills to fuel her passion

Figure 4.1 A dress modified by Jade for *The Sims*, designed by Jewel Millard

for design and, as we will see later, eventually her understanding of and passion for the power of computers beyond design.

Jade became so proficient at design that the other girls in the club began asking her to make custom clothing for their Sims. When the girls went to give presentations at a conference on games and learning, all of the girls' PowerPoint presentations included Jade's designs, ranging from a child's room decorated with a *Sponge Bob* bedspread and tabletop to a bridesmaid dress. For the presentation, Jade created a Sim for each girl and created custom clothing for each one.

Jade's proactive approach to learning to be a designer came to characterize her learning in the club as a whole. She was highly motivated to accomplish her own goals. For example, when Jade was driven to learn how to download content from the extended *Sims* fan community, her major obstacle was not knowing how to deal with compressed files. During one session in the club, she learned how to use WinZip and WinRAR, and this unleashed a downloading frenzy. A short time after that session she had over eight hundred downloads on her computer at home. Eventually, as we see later, she began to upload her own designs to the Internet and let people around the world download them. Again, given a little initial fuel, she drove her own engine and learned to fuel it herself.

Once Jade learned a skill, she readily transferred it to new areas. For example, after she learned how to create clothes, she saw Beth creating wallpaper and flooring, which sparked her interested in learning those skills. Watching Beth create several types of wallpaper styles was all the information she needed to go home, apply the techniques she learned for clothing creation, and make creative wallpaper. Here is what she had to say: "I did nursery wallpaper and you go and you select what you think you want to start with . . . and then I take it into Photoshop document, which is the same as I use for the clothes and I pretty much do the same steps that you do for making clothes . . . you have to select the object and paste it over and merge it down to get the image to actually show up in the game . . . you have to import it in order for it to actually show up in the game."

Jade and School

Jade did not just learn how to design. She learned how to learn and how to be a proactive learner. But did Jade learn to be a proactive learner in school? Yes, and yet this proactive approach did not work. Her experience with design motivated her to learn more about computers, which she came to see as "a source of power." When she went to high school, she wanted to take programming and deepen her computer skills. But she was barred from the programming class because her earlier math grades did not reach the minimum required for the course. Of course, programming might well have given her a context within which to understand mathematical thinking as useful and important.

She went on to take a course in graphic design. However, the course covered what she already knew and the teacher would not let her move on without endlessly repeating skills that had by now become second nature for her. She longed for more challenge and the opportunity to sharpen her skills. Her desires and skills went unrecognized. The institution would not see that she had become a different person. Her school saw learning to be about knowing, not being. The school did not realize that becoming someone, gaining a new identity and sense of self, is the foundation for new knowing.

Jade and Her Father

Real learning is life changing. This is so because real learning is learning to be, to take on a new identity in life, and not just learning to know or do. Jade's learning transformed her and, in the act, transformed her relationships as well. We can see this clearly in her relationship to her father.

Jade always had a close relationship with her father and, as a result, experienced a great deal of difficulty while he was stationed overseas in Iraq for a year. When he returned she found that it difficult to talk with him. He appeared distant and angry. One thing that began to bring them together again was Jade's growing interest in computers. Her father had been the one in the family who took care of the computer and made sure that Jade and her sister had what they needed to complete their homework.

After Jade presented at a conference on games and learning, Jade's father began to take a real interest in her participation in the TSG Club. During this time, Jade's learning was progressing exponentially, and she received a great deal of praise from the conference participants. She was beginning to see that she was interested not only in design but also more generally in computers because as she said "computers give you power."

Jade's father took note of her growing interest in the power of computers. He encouraged her and applauded her accomplishments. In fact, Jade had this to say about her father: "I was talking to my dad about [my interest in computers] and he supports me going to college for computer programming or something like that. He didn't really support what I wanted to do before. I wanted to go into interior design but since I've been doing all of this stuff on the computer it's made me realize how much I like working on the computer." When asked for additional information about what she and her father talked about, she replied, "I was talking to him and people say I'm really good on the computer, so I told him I was thinking about taking as many classes as I could towards this. And he thought that'd be really good. He said 'That's where all the money in the future is going to be.'"

Although Jade came into TSG with access to a computer and high-speed Internet, she did not have a computer of her own. The computer at home belonged to her family. When Jade was preparing for the conference, she started talking about wanting her own computer. Her father was well connected in the community, and, after spreading the word that he was looking for a used computer for his daughter, a neighbor offered to sell him a computer fully equipped with software for one hundred dollars. This was a bargain that Jade's dad could not pass up, even though they suspected that the computer would need serious upgrading.

Coincidentally, the neighbor had an associate's degree in computing and was eager to help Jade upgrade the computer. He sat down with her and explained the components of the computer and what would need to be upgraded. Her father bought the necessary parts and had them installed. Once Jade had her new computer, she set off on a wild binge to create new *Sims* content.

In addition to the computer, the neighbor gave Jade a flash drive and explained its capacities. After the conference, Jade bought several other flash drives and taught the other girls how to use them for transporting downloads and content from one computer to another. Soon she wanted a brand new, state-of-the-art laptop, which her father finally purchased for her. Shortly after receiving her laptop she commented, "If it wasn't for me being in this program, my dad would have never spent the money to buy me a computer like this!"

This story shows us one way in which learning for Jade began to change how people related to her and she related to them, including the central relationship she had to her father. In turn, her changed relationship with her father began to change her attitudes to school. Now she saw a need for school. She wanted to take programming and delve deeper into computers. It is, of course, deeply unfortunate that the school was unable to see the new Jade and adapt to her, unlike her father.

Jade transformed her interest in design into a general interest in the power of computers. This transformation located her interest in a context that was more appealing to her father and, indeed, to many men. Her interests now seemed less gendered as "female." But a crucial question arises. If Jade can learn to talk about her interests in ways that allow males to recognize and admire them, can males, school, and society learn to see her interests in design and *The Sims* as not inherently gendered as female though often associated with females? Can they see such interests as crucial for the development of twenty-first-century skills important to society as a whole? We will see later that thinking like a designer and being able to build and learn from simulations are, indeed, crucial twenty-first-century skills.

The Sims Community

A major reason Jade wanted to create her own clothes is that she saw what other players around the world were creating and uploading to fan sites. She saw there was a global community of *Sims* designers, and she wanted to be part of it. She started to upload and display her designs on EA's *Sims 2* fan site (http://thesims2.ea.com).

Betty and Beth created a group Sim Page on the fan site, http:// TheSims2.com for the girls in the club, since not all of the parents wanted their girls to have an individual Web presence, no matter how well protected. Jade's mother did not monitor Jade's online activity as much as some other mothers did, so Jade created her own Sim Page. She didn't add any personal profile information, perhaps because of the general worry the girls had picked up from their parents about Internet stalkers.

Jade uploaded a *Sims* video that she created as well as custom-designed Sims. She quickly got lots of feedback from other players who downloaded her content. The comments were uniformly positive:

- "I love your sim 'Andy.' I think she's really pretty. All your other sims are cool too. I rated it 5 stars. I would rate it 10 if I could, lol."
- "I just downloaded Lisa the sim you uploaded, I LOVE her pajamas!! You did an awesome job!"
- "Hey, Your Sims Are Really Cool. I've Downloaded 'Andy' & I Rated Her 5 Stars. =] I've Seen Your Others & You Make Awesome Sims! Keep Up The Good Work!"[1]

Jade also received requests from other players:

you do good work

I was wondering if you do uniforms I am doing military themes if so that would be great and keep up the good work.

I luv the pjs

I think you have done awesomely with your sims, did you make them your self? if so can u make a pair for me sayin: no dancing on this surface! thanx plz can you sign my guestbook to know you got the message thanx.

The last request referred to a series of pajamas that Jade created with different themes, such as "hot pajamas: these super cute lite pink pajamas have a cute cookie graphic on them and say 'one tough cookie.'"

Jade developed an international audience of several hundred people who took her creations, praised her, and gave her feedback that improved her work. A girl from a rural area, a girl who had few connections outside her local area, had become a global citizen. She was connected to the wider world. Her new identity allowed this, but these connections furthered her identity transformation.

Jade could have stayed in this worldwide *Sims* community and moved up in status in the community. Many girls and women have. We will see some of them in later chapters. But Jade found another community that allowed her to expand her horizons even more.

Jade and *Teen Second Life*

After the girls in the TSG Club had played *The Sims* a good deal, Betty and Beth, along with a third woman, Barbara, introduced the girls to *Second Life*. *Second Life* is a 3D virtual world like the internationally renowned game *World of Warcraft*, where hundreds or thousands of real people can be in the virtual world interacting together at the same time. However, *Second Life* is not a game, but a world built and sustained by its participants. *Second Life* includes its own building tools, and people have built a massive array of different environments in which participants can interact with each other (see *Second Life*'s official Web site at http://secondlife.com for more information). Players in *Second Life* (*SL*) can own and develop their own land. They can build things if they master the building tools, or they can buy things other people have built. Building in *SL* is a type of modding. Whatever a player builds in *SL*, whether it is a house, clothes, buildings, vehicles, objects, or even games, is owned by the player, who can either give it away free to others or sell it for Linden dollars, the currency of *SL*. Linden dollars are legally tradable for U.S. dollars, so people can make real money in *SL*, and many do. (*Second Life* was created and is maintained by the company Linden Lab, and the currency is named after this company.) Betty, Beth, and Barbara wanted the girls to progress from the single player *Sims* game to a safe virtual world where they could interact with other players. There was a teen version of *SL*, open only to teenagers and with more restrictions on behavior and building than in the adult version. Adults could enter *Teen Second Life* (*TSL*) only after undergoing a background check and, even then, were restricted to private islands, mostly ones that had an educational purpose (see http://teen.secondlife.com/). Betty, Beth, and Barbara purchased their own island, TechSavvy Isle, that was initially only accessible to the TSG Club participants and adult mentors, though later it was opened to other teens. The girls could visit other islands that were part of *TSL*.

Betty, Beth, and Barbara initially asked the girls to explore the teen world and take snapshots of things they found interesting. Jade quite quickly replaced her standard avatar with a customized avatar "skin" that she purchased (the girls had been given some Linden dollars to spend). This avatar was tall and thin, with long dark hair (unlike Jade who was relatively short and plump, though she does have dark hair).

Jade also found a number of animations, including one that made her avatar walk and pose like a runway model. (There are many types of animations in *SL* and *TSL*, for objects as well as avatars.) At one point, Jade found an announcement of a modeling event in *TSL* and participated as a runway model (she found the model animation through connections with the other teens involved in this event).

One of the girls' first building projects was to create their own home on the island. They also helped create a welcome area for visitors and other buildings. Jade became quite good at finding items created by other teens and eventually start creating them herself. For example, she constructed a grill with hamburgers cooking on it (including an animation of smoke rising from the grill).

As the girls engaged with designing their own island, some of them became intimidated by the quality of the content created by other teens, content they had seen on their forays around the *TSL* world. This was not a concern that Jade shared. In fact, she rented a space in a shopping mall and proceeded to build her own store where she sold her own creations. She had developed a confidence in her content-creation abilities through her previous experience with content creation for *The Sims*. Her work with *The Sims* served as good preparation for future learning for her work in *TSL*. Her identity transformation continued: she was now an entrepreneur with her own store.

Jade learned how to create many things in *TSL*, ranging from clothes to furniture. Some of this content was created by modifying existing objects or clothes provided by Linden Lab, using the in-world object editor. She also used Photoshop to edit clothes and import them into *TSL*, similar to what she did with *The Sims*.

The *SL* in-world object editor (a 3D building tool) is an intimidating graphical interface that allows players to manipulate geometric primitives (called "prims") by stretching them into new shapes, changing their texture and physical qualities, and linking them to other prims. Everything in *SL* and *TSL* is built with prims. The object editor provides numeric information about the object, including size, rotation, and position, each in 3D. Objects are manipulated in the world with a set of 3D geometric coordinates.

Jade's travels in *TSL*, far from intimidating her, made her desperately want to open her own store to sell what she created. Since initially the club's island was not open to the public, to avoid vandalism and to give the girls a private space to learn skills, Jade used her own babysitting money to buy Linden dollars so she could rent space in the shopping mall we mentioned previously. She learned how to use the currency-exchange tool, as well as how to locate an appropriate space and pay rent. She decorated the store herself and sold furniture as well as clothing. She had to learn how to set objects for sale, price them, and restrict modification or copying by other users. Jade's knowledge of copyright norms in *TSL* led her to intervene when another girl tried to label and sell content made by another teen as her own.

TSL was but one point on Jade's trajectory of learning and identity transformation. She is still very much on that trajectory. Notably, even by the time she was a storekeeper and business woman in *TSL*, Jade did not think of herself as an expert. Despite her dramatic growth in technical skills, she still rejected the label "expert." When someone identified her as the "group expert" in the club, she said, "I don't know about expert . . . I've always helped people when it comes to different stuff." In fact, she actually preferred to be seen as a "helper." Asked what she thought her role in the club was, she replied, "I help everybody. That's been my job."

Jade as a Pro-Am Prosumer

Jade is an example of an important phenomenon today. She is becoming what has been called a "pro-am" (a professional amateur; Leadbeater & Miller, 2004). Digital worlds and tools are already used to massively enhance learning at earlier and earlier ages, especially out of school. According to Leadbeater and Miller, we live in the age of pro-ams. Many of these pro-ams are young people who use the Internet, communication media, digital tools, and membership in often virtual (sometimes real) communities of practice to develop technical expertise in a plethora of different areas. These include video games, digital storytelling, machinima, fan fiction, history and civilization simulations, music, graphic art, political commentary, robotics, anime, fashion design, and many others.

These pro-ams have passion and go deeply, rather than widely, into their interests. In any field, developing such a passion is a sine qua non of deep learning that leads to expertise. At the same time, pro-ams are often adept at pooling their skills and knowledge with other pro-ams to accomplish bigger tasks or to solve larger problems. These are people who do not necessarily know what everyone else knows but do know how to collaborate with other pro-ams to put knowledge to work to fulfill their intellectual and social passions.

Passion, as we have seen, drives practice, and practice drives expertise. There is one fascinating and ironic thing about Jade's passion. We talked in an earlier chapter about "cultivated children." These are the children, often from privileged homes, whose every moment of free time is scheduled by their parents for activities that will accelerate them in school and society. Jade was not a cultivated child. Her time was considerably less structured. When she found her passion, she had time to go deeply into it—and there is no other way than going deep if you want mastery. Too many of our over-scheduled high achievers may not have enough time to find their passion and go deeply into it. It would be a shame if they were only good at school-based learning, especially in schools where no twenty-first-century skills are offered.

At the same time, the pro-am phenomenon raises deep questions about equity in the other direction. Since most of this learning takes place outside of school today, as it did for Jade, it is often easier for more privileged young people to gain the digital resources, mentoring, and access necessary for becoming pro-ams. We do not know how pervasive the pro-am phenomenon is among less-privileged young people, but many community programs seek to offer less-privileged children the opportunity to engage with digital communities of practice to become pro-ams and, in the process, pick up important twenty-first-century skills. This is, of course, what happened for Jade.

Jade is also a classic example of what the futurists Alvin and Heidi Toffler (2006) have called a "prosumer," that is, a consumer who produces and transforms and does not just passively consume. At first, prosumers engage in productive activity off-market for status and not money, and they do so as part of a community of like-minded pro-ams. However, as the Tofflers point out, such prosumer activity often eventually impacts markets when people like Jade ultimately sell their goods or services. The Tofflers believe such activity, though unmeasured by economists, is a big part of the global economy and will become a bigger part in the future.

Jade: Typically Untypical

Readers now might ask, "How typical is Jade's story?" The answer is, ironically, it is typical in being untypical. We live in a new world. School is about bell curves and bell curves are about being typical, being normal, and being in the middle. However, in developed countries, faced with a global competition, worldwide surplus labor, and low-cost service centers like India and China, the economy does not and will not reward people who are simply "typical."

Three fifths of the jobs in the economy of a developed country are service jobs or the remaining manufacturing and manual labor jobs (Reich, 1992, 2001). Many of these jobs neither pay well nor carry generous benefits. One fifth of jobs involve technical work (e.g., putting together electronic circuit boards or carrying out routine medical tests) that is not always high paying. The remaining one fifth of jobs involve producing and designing new knowledge, products, and services, or otherwise engaging in innovation. These jobs are often highly rewarding in pay, status, and ego satisfaction.

Being typical, that is, having the skills lots of other people have, does not necessarily lead to success. Plenty of college graduates with high-tech skills face stiff competition from fellow graduates with equal skills. Success in the modern world is risky even for those who are good at school (witness, for example, the incredible competition these days to get to elite colleges).

More and more success requires three things: passion, persistence—the combination of which we have called "grit"—and the ability to innovate. Success today requires deep mastery and mastery that goes beyond standard skills to the ability to innovate. Such mastery requires lots of practice. Research in cognitive science, the science that studies the human mind, tells us that real expertise in nearly anything requires about ten years or ten thousand hours of practice (Ericsson, Krampe, & Tesch-Romer, 1993; Gladwell, 2008). Such commitment requires passion and persistence (i.e., grit) plus, of course, opportunity.

Even passion, persistence, practice, and the ability to innovate are not enough. More and more, success will require people to collaborate as members of what we might call "smart teams" and as members of what we call in Chapter 6 "passionate affinity groups." In smart teams, sometimes called "cross-functional teams," which are common in today's high-tech workplaces (see, for example, Parker, 2002), each member has deep expertise as well as the ability to integrate this expertise with the different specializations of every other team member. Each team member can go deep, but each can still see the big picture. Smart teams operate in such a way that the team is smarter than the smartest person in it (the opposite of a "committee").

Though she learned a lot about collaboration thanks to her interactions with others in *The Sims* and *Second Life* communities, Jade did not have much opportunity to learn how to be a member of a smart team. Such learning would have involved collaboration and co-design on a project that required different, but related, forms of specialization. Different girls in the club would have to develop different passions leading to different specializations and then engage with a project that would recruit and require these different specializations and their integration.

Passionate affinity groups are communities of people, and there are now many on the Internet, who organize around their shared passion and not on the basis of race, gender, class, ability, or power. Such groups are often not age graded: young, old, and in between are all present together. Newcomers are not tightly segregated from old hands and experts, and these old hands might be young or old. Passionate affinity groups provide lots of mentorship, opportunities for everyone to become a mentor, and multiple ways of contributing. Different people lead or follow on different days or in different activities. Standards are high, but access is open to anyone with passion and willing to abide by the community's rules and norms.

Jade first entered a global passionate affinity group committed to designing for *The Sims*. This community nourished her new sense of self as a designer and served as a bridge to a passionate affinity group devoted to building in *Second Life*. This community, in turn, nourished her new sense of self as a designer, as a growing expert with computers, and as an entrepreneur.

Passionate affinity groups operate by what has been called the Pareto Principle or the "80/20 principle" (Shirky, 2008; see Chapter 6 in this volume). Twenty percent of the people in such a community make 80 percent of the contributions and 80 percent make 20 percent of the contributions. This distribution ensures that everyone's contribution, no matter how uncommon, counts, while fully recruiting the accomplishments of the top performers, the people with the most passion. This is very different from school, where the bell curve reigns, where nearly everyone is expected to do the same things and activities are mostly designed to recruit "average" abilities. At the same time, for success in real life, many people will have to find the passion and the community that leads them to be in that top 20 percent producing 80 percent of the content.

The nature of our modern world means that many people cannot gain high status on the market, in terms of how much money they make and how much status they get from the work they do for pay. Remember, only about one fifth of workers in a developed country get high rewards and high status for what they do on the market. A great many people will need routes to status and a feeling of worth off market from what they do for passion, not for pay. A great many people are already finding this status and worth off market in passionate affinity groups on the Internet.

As we mentioned above, many people today who seek status off market for what they design and produce as part of passionate affinity groups eventually go on market and sell their products. Ironically the search for status off market leads to innovations on market and has an impact on the national and global economy. This impact is already large and will be yet larger in the future.

When we think about Jade, we see some twenty-first-century skills or, better put, twenty-first-century ways of being: passion, persistence, practice, collaboration, passion communities, innovation, off-market status, innovation, and "prosumption" that eventually leads to expanding markets. The people who master this twenty-first-century way of being will not be typical. Each will have to be special in terms of sharing their deep expertise and passion with some niche community, not the masses. For success, ironically, a great many people will have to be untypical.

Jade raises a final question. She was developing a twenty-first-century century mind-set and array of skills, dispositions, and attitudes, but she was not successful in school. Her school failed to acknowledge, recruit, or enhance her new passion for design and computing. There is even the danger that school will eventually kill her passion. Jade found her passion and a new self out of school, as are so many other young people today. Until we make radical changes in our schools, this out-of-school learning will be the main curriculum for twenty-first-century skills. School will become less and less relevant, if not actually detrimental, for many "atypical" young people.

Chapter 5

How Passion Grows

A Retired Shut-In Goes from Making a Purple Potty to Gaining Millions of Fans

Passion

This book is about understanding large social and educational issues through the lenses of specific stories of creative girls and women. School, as we know it, is usually about learning information—such things as facts, rules, and principles. Today we regularly base tests on knowledge of such information and hold schools accountable for failing test scores. There are lots of problems with this model.

First, there is plenty of evidence that people who just learn information, in science, for example, cannot use this information to solve problems (diSessa, 2000; Fair Test, 2007; Gee, 2004). They may be able to write the information down on a test and even get good grades, but they cannot solve real problems. As we discussed in Chapter 2, in one study, college students taking physics could not correctly say how many forces are acting on a coin when it is thrown in the air, despite the fact that the answer can be deduced from Newton's laws of motion, and many of the students could write these down on a test (Chi, Feltovich, & Glaser, 1981).

Second, information, which was once hard to get, is now easy to get. Students can find a good deal of what the teacher is telling them in the classroom on the Internet. They can even fact-check their teachers. In an age where information is cheap, it becomes crucial to be able to sort through it, to make good decisions about what information to trust, and to be able to use it for good purposes. We are still making our students information storage devices in an age of near-infinite storage via digital technologies.

Third, memorizing information does not lead to the ability to be innovative and creative. We have seen already that, in a developed country, standard skills—being able to do what lots of other people can do in the same way others can do it—will not bring success (Friedman, 2005). Workers can now be trained to use these skills and apply them in jobs carried out in low-cost employment centers around the globe. Many standard tasks are now automated and carried out by computers and other technologies monitored by low-level workers. These trends will only increase in the future.

Learning for success today has to involve deep mastery of problem solving and the ability to innovate. Such mastery does not come without a great deal of persistence and practice. Few people will put in such and persistence and practice unless they have a passion for what they are doing. We earlier defined what we believe is a core twenty-first-century disposition: grit (Duckworth et al., 2007), which we defined as persistence plus passion.

We need to know how people gain passions because passion is the fuel that motivates effort, practice, and mastery. There is a further complication here: today, it is not just young people who need to gain a passion for learning. People of all ages will need this passion in a quickly changing world that readily makes old information, skills, and even old jobs obsolete. They will also need this passion if they are to have satisfying lives in a world that cannot give everyone status and a feeling of control based on the status of their job and the money they make.

The Purple-Potty Theory of Passion

Where do passions start? Our answer is that passions often start with little things, not big ones. For example, let us tell you about a passion that started with a purple potty. This passion did not start early, but late, in life. Meet Tabby Lou, a woman who found a passion and fame after she retired.

Tabby Lou is a sixty-one-year-old single woman living in a rural area of Pennsylvania. We interviewed Tabby Lou, along with a number of other adult women, about her *Sims*-related learning (at that point, *The Sims 3* had not yet been released, so all of the women were using *The Sims 2*;

Hayes, King, & Lammers, 2008). She is a former computer instructor for the U.S. Postal Service. At the end of her career, she served for fifteen years as a postmaster until she retired in 2003 due to a health condition. Her health problems have made her homebound. In fact, she only leaves her home twice a year. Tabby Lou is most certainly not the sort of person who figures in discussions about school reform and the future of learning. But times have changed, and we argue that she is just as relevant to lifelong learning in the twenty-first century as a kindergartener entering school for the first time.

Tabby Lou's adult daughter bought a copy of *The Sims* when it first was released in 2000 and, watching her daughter play, Tabby Lou became interested. By the time *The Sims 2* was released, her granddaughters also loved to play the game. Tabby Lou got two computers so the girls could play when they came to her house. At that point Tabby Lou just played the game as part of a three-generation family of *Sims* fans.

One day, one of her granddaughters told Tabby Lou she wanted a purple potty to put into her Sim houses. The game does not come with purple potties. But what grandmother would disappoint her grandchild? Tabby Lou decided to build one for her. That meant she had to learn to make content and to recolor content for *The Sims*. She had to become a designer and not just a player. This took lots of effort, effort fueled at this point by her passion for her grandchildren. Now Tabby Lou had grit.

Today, only a few years later, Tabby Lou is an internationally known and respected designer for *The Sims*. At the time we wrote this chapter, her creations had been downloaded more than thirteen million times. Think about this for a moment. What would you think of a fashion designer who had sold thirteen million designs? What about an author whose books have been read thirteen million times? An ad campaign that got thirteen million responses? An entertainer that sold thirteen million tickets? Tabby Lou is sixty-one and infirm, but she is a rock star in her world. And it all started with a purple potty: "I became involved because my Granddaughter wanted a purple potty. Yep, just because of a little purple potty request from her. I read everything I could get my hands on and went from there."[1]

It was not just a purple potty, really. It was a granddaughter who wanted a purple potty. And it was a new digital world that had tools that would allow a sixty-one-year-old grandmother to build one. But learning to use the tools requires grit.

Here is what we might call a purple-potty theory of how passion develops. First, there is something you really want to do. In Tabby Lou's case, it was to fulfill a granddaughter's wish. You can do this thing, and you badly want to, but only if you learn something deep (for Tabby Lou, creating content in *The Sims*). Good tools are available that allow you to learn it, but the learning requires persistence. You care enough to persist. Now you have grit. You are launched.

Next, the learning "hooks you" in its own right, as learning to be a designer captivated Tabby Lou. The learning inspires more passion both for the subject and for the identity associated with being knowledgeable about this subject. Tabby Lou transferred her grit in service of her granddaughter to grit in service of design. Eventually, she became famous and her granddaughters are proud of her: "My granddaughters think it is fantastic to have a Grandma that does recolors." Currently, she is a featured artist on a major fan Web site, and she is internationally known in *Sims* design communities.

What starts a passion need not seem big or highly significant. Purple potties will do. The trajectory is as follows: have a strong desire to do something ‡ find the needed tools ‡ gain grit in the service of doing it ‡ get hooked on the learning ‡ transfer grit to the learning itself ‡ become successful. Yes, we live in a world where anyone can become successful in a high-tech area, including a sixty-one-year-old shut-in, shut in at home, but not shut out of the globalized world.

Tabby Lou

What does it mean, really, to get hooked on the learning? Why didn't Tabby Lou just quit after she made the purple potty? When she chose to continue, why didn't she just design for her granddaughters? What made her go global and share her creations with an international audience?

One answer to these questions is that Tabby Lou was confined to her home. Here is what she said in one interview: "It makes my day go a whole lot faster. When you do not go outside due to health reasons, all you can do is look at the walls. I do read sometimes, but this is a whole lot more enjoyable and rewarding."

It is easy to miss how modern and new this all is. Tabby Lou's life changed radically. She got sick, had to retire, and could not leave home. In the past, she would have been both shut in and shut out from the world. Today, she can start a new life by learning something new deeply. The tools are there—we will see in a moment that a whole community is there to help—and she just has to add grit. It is like adding water to flour, along with a little yeast: you get dough.

Tabby Lou is not interested in making money. She is typical of many people today, young and old, who are gaining status off-market (Toffler & Toffler, 2006). She does what she does for a community of people all over the world not for profit. But if she wanted to, she could make money and sell her designs. If even a small fraction of her thirteen million download-ers paid her a dollar, she would be rich.

More and more in the twenty-first century, people will have to start over at all different ages. Tabby Lou started over at sixty-one, in retirement and in poor health. As she seemed to disappear into her home, far from becom-ing invisible, she became highly visible as a person with a global audience.

Starting over and gaining status off-market when her life looked like it had taken a turn for the worse was not Tabby Lou's only motivation. What really hooked her was the community she could enter and serve. She did not want just status off-market, but she wanted to help people and give them pleasure through her creations. This is, for many people, a much deeper motivator than profit.

There is a big community of *Sims* designers in the world. They welcome people like Tabby Lou, no matter what their age, race, creed, or disability, as long and only so long as they have a passion for *The Sims*. They offer people like Tabby Lou a myriad of online tutorials they have designed and all the mentoring they can ask for. Better yet, they also offer people like Tabby Lou the opportunity to help, serve, and mentor others.

Tabby Lou is part of two fan communities: The Sims Resource (TSR) and Mod The Sims 2 (MTS2). Tabby Lou readily acknowledges the help she has received from *The Sims* community along with each item she creates. Each item Tabby Lou uploads to fan sites features "credits" that acknowledge people who have helped her. First she credits the person who requested the recolor, the person she is serving. Then she credits others in the community who have helped her. For example, for one of her items, she credits the creators of SimPE, a software mod featured on MTS2, an art instructor at MTS2, and an artist at TSR. Tabby Lou is part of a professional community (really a pro-am community). She has colleagues aplenty.

Even as an expert, Tabby Lou is still getting help while always giving it. People enter the community knowing they will sometimes receive help, get mentored, and follow the lead of others and sometimes give help, mentor others, and lead themselves. Newcomers can expect, if they work hard enough, to become helpers and mentors themselves. Experts can expect to be able to continue learning and growing, pushed by the support of the community.

Tabby Lou is driven by a passion not just for design but for her fans and the community. She says that "my fans and their comments and requests have a tremendous amount of influence in what I do in regards to recolors." She elaborates further: "It has given me great satisfaction and fulfillment that people like my recolors. These feelings make my days less lonely and boring, even if it is only from a recolor of an object in a computer simulation game. I LOVE all of my followers, old and new. Without them, I would be a very lonely person." TSR has a system that allows fans to thank *Sim* creators for their content. Tabby Lou has a guest book where people can leave her messages. As of August 2009, she had been thanked more than one million times. Talk about motivation; this appreciation sure could give you a sense of belonging and a feeling of self-worth. Profit begins to pale as a motive, especially profit with no thanks.

The Sims communities of which Tabby Lou was a part use praise—they readily offer encouragement and support to each other—not harsh criticism or "flames" to help people learn. Often, Internet forum boards feature flames, or instances of ridicule, dismissal, and blunt criticism. We may be surprised to find that *The Sims* communities of the sort Tabby Lou belonged to do not engage in this behavior. We could assume this is because these communities are made up of a majority of women. But, in our view, the real reason is because they are true professionals. They have a passion for what they do and want to see it improve and spread. They want to see their community grow, not shrink. They are often mutual fans, admiring the people who also admire them.

Now we have to add to our purple-potty theory of passion. Earlier we argued that the trajectory of passion is have a strong desire to do something

‡ find the needed tools ‡ gain grit in the service of doing it ‡ get hooked on the learning ‡ transfer grit to the learning itself ‡ become successful. Now we have seen that becoming truly hooked on learning often requires two things.

First, you need to be willing and able to change. You have to be willing to start anew, to become somebody new, and to take on a new identity. Tabby Lou could have viewed herself just as a retired, shut-in grandmother. Instead, she came to view herself also as a designer (for a video game), as a mentor and helper, and as an international contributor. This is not what we typically associate with retired, shut-in grandmothers. The world has changed. Like Jade, Tabby Lou is typically untypical.

Second, it helps to have a community behind you. It helps even more when that community allows people to help and be helped, to mentor and be mentored, to lead and be led. It helps, too, when that community is a community of true professionals (pro-ams in this case) who use praise, support, and encouragement to spread their passion.

Here is our amended, expanded, purple-potty theory of passion. The new trajectory of passion is: have a strong desire to do something ‡ identity and community ‡find the needed tools ‡ gain grit in the service of doing it ‡ identity and community ‡get hooked on the learning ‡ transfer grit to the learning itself ‡ become successful. Note we have put identity and community into the trajectory at two points.

What happened in Tabby Lou's case was this: She had a need she cared badly about filling (making a purple potty for her granddaughter). She found out there were good tools available to fulfill that need, but they were not easy to learn. At that point she found that *The Sims* community had resources that would help her, but she also knew she had to be willing to see herself as a budding designer (if only of purple potties). She was now at the periphery of the community. She needed their tools and their resources. She needed them so she could help her grandchild.

Tabby worked hard to fill the need and make the purple potty, but her changing identity and her exposure to the *Sims* community made her want more and hooked her on the learning that would solidify her new identity as a designer and her membership in a community that brought her status, support, and friendship. That meant much more hard work. But this was not a problem because Tabby Lou had grit. It meant, too, later, that she would receive millions of downloads and hundreds of thousands of thank-yous.

This book is about women and video games. Tabby Lou does not consider herself a gamer. When asked if she was a gamer, she said, "I do not consider myself a gamer. I do not play the game. I have more fun in creating recolors. I do have a lot of computer games that I used to play before I got into this recoloring phase. I enjoy being a creator more than being a gamer."

Tabby Lou started as a player but eventually became a designer of content for the video game she had played. Such people are often called "game designers." *The Sims* was designed by the game designer Will Wright, whose former company, Maxis, publishes *The Sims* games. Tabby Lou works for Will Wright. She does not know him and she does not get paid, but she is building content that appears millions of times in Wright's game. Wright realized that he could put people like Tabby Lou to work, free of charge, saving himself money and changing their lives in the act (see, for example, Kosak, 2005; Wright, 2006). Wright, without a doubt, already knew our purple-potty theory of passion. Now, as we've heard personally from many game designers, Tabby Lou does not play much, she is too busy designing.

At a deeper level, though, as we have seen when we discussed modding, playing games and designing them are closely related acts. To play a game well, you have come to understand its design well enough to use its features for your own goals and success. This understanding can lead to a desire not just to play the game but to make it, too—to mod it or build content for it. Good game designers know that it is in their interest to feed this desire. This not only produces games like the famous mod that turned into *Counter-Strike*, made from the game *Half-Life*, played worldwide, but also produces people like Tabby Lou.

Preparation for Future Learning

Readers may well have an objection to Tabby Lou's story. Tabby Lou did computer training for the U. S. Postal Service before she became postmaster. Surely, then, she was well prepared for learning to design for *The Sims*. Tabby Lou readily acknowledges that nonetheless, she still did a lot of learning to attain her level of expertise. But there is an important issue here: the notion of "preparation for learning" (Bransford & Schwartz, 1999) is crucially important.

Deep learning occurs when people find a passion. But passion is not enough. Persistence is required as well. Deep learning requires grit, which we have said is passion plus persistence. However, when we look at any complex skill, we can ask, what earlier experiences are good preparation for learning this skill? Surely earlier experiences with computers, art, and design are likely to be good preparation for learning to design for *The Sims*.

We also can ask, what other things does learning one skill prepare a person to learn in the future? Learning anything deeply creates possibilities for learning other related things in the future and invites a person to move along one of several learning and growth trajectories. The person is well prepared for some things and not for others. Certain doors into the future are now at least ajar, if not wide open. What does learning to design for *The Sims* prepare one to learn in the future?

We can also ask, what helps learners who come to a deep learning task without good earlier preparation? What happens when people find a passion for learning for which they are not particularly well prepared? How can they gain grit?

Why should we care about this issue of preparation for future learning? We tend to think of learning as something done in the here and now, with all the action in the present: what does this learner know right now? The way we assess learning in school tends to encourage this point of view. But learning anything, at least anything deeply, always creates a history reaching far back into the past and extending out into the future. We should think of learning in terms of learning trajectories, or paths of development.

It takes a long time—in many cases years or thousands hours of practice, as we have noted earlier—to get really good at something difficult. When we test knowledge and performance in school, we are often testing people who have had very different amounts of practice. Take these hypothetical children as examples: Young Janie comes from a home where, long before she went to school, she had many hours of practice in tasks that are good preparation for future learning to read (e.g., being read to, playing with language and with letters, writing with invented spelling). By fourth grade, she already has read a great deal, at home and at school. Young Johnnie has had much less of this preparation.

Janie and Johnnie take the same reading test on the same day, and Janie performs a lot better than Johnnie. However, when (and if) Johnnie gets the same hours of practice that Janie has already had, he may be just as good or even better than Janie is now. But if we tell Johnnie, because of the test results, that he is a poor or slow learner, he may give up. He may come to see himself as dumb rather than just poorly prepared and not well practiced.

The test should have told us where Janie and Johnnie were on a trajectory of learning to read. It should have told us that Johnnie is doing just fine for someone who had much less preparation and practice than Janie. We should have been planning how to get Johnnie many more hours of preparation and practice without demeaning him or boring him to death with skill-and-drill teaching methods. The way to do that might well be to help Johnnie find a passion that demands reading.

Preparation for future learning is important, too, in our world where things change so fast that we cannot rely on our current knowledge and skills. Things we have mastered will become obsolete or we will outgrow them as our passions transform. To remain vital, people will have to think about how their learning today is good preparation for a lifetime trajectory of learning (Partnership for 21st Century Skills, 2007).

As we described earlier, the science educator Andy diSessa (2000) talked about how the tinkering he did as a boy in his garage with his father was good preparation for the future learning he did in college when he took physics. The tinkering was good preparation not because of any physics facts it taught him. In fact, it did not teach him much physics directly. It was good preparation for learning physics because of the attitude it gave him: technical learning was not hard, it was the sort of thing he could do, and it was rewarding to know how things worked.

In contrast to Andy diSessa's experience, compare children whose parents decide they do not have an aptitude for math long before the child has done much earnest practice in mathematics. Early struggles with mathematics, sometimes abetted by poor math instruction in school, leads to this attitude. The child and his family have often simply not had earlier experiences that make them see mathematics as both learnable and valuable.

However, children faced with learning *Yu-Gi-Oh*, for instance—a highly complex anime card game involving reading and math—rarely have the same attitude. They gladly struggle and learn even if they have not had any good earlier preparation for learning the game (though, of course, many have). Why? *Yu-Gi-Oh* often triggers their passion. But they still have to persist past initial struggles. We will see later that community has a lot to do with how and why many young people can display grit for learning *Yu-Gi-Oh* and not school mathematics. Like *The Sims*, *Yu-Gi-Oh* is associated with a massive, global Internet community that offers support of many different kinds (see, for example, Konami, 2008; Pojo, 2009; Yu-Gi-Oh Wikia, n.d.).

Izazu

Tabby Lou's computer background was good preparation for learning to design for *The Sims*. Next we will discuss a woman who did not have such good early preparation for future learning as a *Sims* designer, and consider how she managed to succeed even without such earlier preparation. What was it that made up for her lack of earlier preparation? Why didn't things work as they often do in school, where people without good earlier preparation are seen as slow and behind.

Izazu (her screen name) was a fifty-three-year-old housewife when we interviewed her. Her husband of twenty-five years, "a wonderful man," had passed away. Her little Yorkshire terrier was "just like a kid" to her. When she started playing and building for *The Sims*, she told us that she "didn't know how to turn a pc on." Izazu has a high school education. There is nothing in her past that seems to have given her good preparation for learning to design for *The Sims*, unlike *Sims* designers we talked to who had backgrounds in art or with computers

Today Izazu rates her tech skills as "intermediate," though we would rate them higher. She has mastered Photo Studio, Photoshop, and Paint Shop, among other software. She has learned how to deal with what she calls "the technical aspects" of computers since her husband died. In fact, her relatives regularly ask her for assistance.

Izazu is confident in her ability to learn new technical skills. She is also a proactive learner with grit. She says, "I am on the PC often and there isn't a single day that goes by that I don't learn something new. I am always willing to learn more and more. When I can't get something to work or something is wrong, I just can't rest until I figure it out. So I think I can handle and I have the drive to learn." Izazu got up to speed pretty quickly. She started designing in 2000 and by 2005 she was named a featured artist on the popular *Sims* fan site TSR. What hooked her originally? How did she find her passion? In 2000, she watched her brother and his sons play *The Sims* and the rest was, as they say, history: "I was amazed with the little people on the monitor! So I decided to try it for myself. I was hooked from the beginning! I loved playing the game but my first love was making the homes for the little people to live in. I then purchased every Expansion Pack that came out for *The Sims 1*. *Sims 2* arrived and the fascination started all over again with the game."

For Tabby Lou, the passion started with a purple potty. For Izazu, it started with wanting to make homes for little people on a monitor. We already pointed out above that the seed that starts a passion need not be large or sound particularly significant. Purple potties and little people will do just fine. We know why Tabby Lou wanted so badly to make a purple potty (because of her granddaughter). We do not know why Izazu cared so much about making homes for the "little people." But all that matters is that she did.

Izazu became "addicted" to playing *The Sims*, but her addiction to playing was completely connected to being addicted to building houses in *The Sims* for the "little people." When *The Sims 2* came out, "building was even more exciting with all the possibilities that the game provided for making homes more realistic."

When she started to build in *The Sims*, Izazu, like many others, turned to a fan site, TSR, for help. There she found people more than willing to help and tutorials that others had made to share with new designers or experienced ones wanting to pick up more skills. These tutorials are quite specialized. People without passion for designing for *The Sims* are not going to put up with them long enough to learn much. Here, for example, is what Izazu has to say about some of the tutorials she used:

> I have used a couple of tutorials which are very handy ones when it comes to building sim homes. I like to use different building techniques so I have a variety of different homes. There is a tutorial on how to place Driveways on slopes (either downward or upward). There is a tutorial on how to make any fencing a banister instead of the regular banisters that Maxis provided with the game. There is a tutorial on making a workable sideways garage, instead of the having

it always straight. And there are tutorials that teach you to build arched bridges with ponds . . . Makes for a nice entry to a castle.

Because some of the homes and community lots she built for *The Sims* "turned out nice," she decided to share them with other members on TSR. This site had already offered her lots of help and mentoring in the form of tutorials, advice, and encouragement from other designers. She decided to give back.

As Izazu worked at her passion, she eventually found that she needed certain colors of wallpaper and floors for a "special home" she was building and could not find anything suitable made by other creators on TSR. She could not use what others had built to finish up her house. She had to create these details herself, at least if she wanted them to meet her standards. Already skilled, she had to learn yet new skills. Indeed, deep expertise requires being able and willing always to learn new skills.

Izazu decided to learn how to make her own walls and floors. She also wanted to use custom artwork and paintings to decorate *Sims* homes, so she started to make these, too. She learned to design almost all aspects of her houses. She began to share more of her creations with others on TSR. Then, after much effort, on September 1, 2005, she was named a featured artist at TSR. Izazu had become a recognized expert without any extensive preparation for future learning in this area that we can find.

Today, Izazu, like Tabby Lou, rarely plays *The Sims*. She spends her time designing, building, and sharing. However, she also points out that "if I had a dollar for all the hours I have spent playing the game, I could buy a round trip ticket to the other side of the world." Houses are her passion: "I am a bit particular too. I do not share a house unless I am completely satisfied with it myself. So from start to finish I would say it takes approx. 3–4 days to finish. Sometimes more if I have to make my own walls and flooring for it . . . And I don't just spend a couple of hours a day either, I spend at least 6–7 hours each day on it. I have to get it just right." Why didn't Izazu's lack of preparation for future learning with computers and design hurt her the way it hurts so many children in school? One key reason is the TSR community. This community functions—in many different ways—very differently from school. For all the women we interviewed, fan communities like TSR played a similar role.

When asked what motivated her to participate in her design activities over time, here is what Izazu had to say: "The *Sims* Community is very large. I think there are so many great creators out there for *The Sims*. I love sharing ideas with them and getting some advice too. But most of all, we learn from each other. And there is so much out there to learn too. And they are a great bunch of people too. I now know people from ALL over the world who enjoy the same interests as myself. You can't beat that!" When Izazu was asked what were the most valuable resources in helping her learn to create content, she once again pointed to the community: "Hmmm, well I must say that other creators have been the most helpful when you need some advice. I do read a lot of tutorials and sometimes the best thing is to just explore and poke around all on your own. But there is always room for improvement and other creators can give you the insight and ideas for even more creativity." Asked if there was any one approach that she found herself doing first when trying to learn new content-creation techniques, once again, we hear about the community: "My number one informational resource is the TSR forums. I have asked so many questions there and in no time, I have an answer. One thing about the *Sims2* Community is, we are all ready and willing to help out anyone who needs help." The *Sims* community is not a one-way street for Izazu. She not only gets, she gives: she goes to TSR both to learn and to teach:

> I can't put a number on how many have contacted me for advice but I can say that when someone does, I do my darnedest to help if I can. Well here is an example: Just yesterday a lady wrote me a PM and said she was having trouble with a house. She said that the basement has water in it. She said it was flooded and she can't use the basement. She wanted to know if I had any suggestions as to what caused this. So I replied to her something to try and if it didn't work after that, I would download the house and put it in my game and try it . . . I told her that some of Sim Neighborhoods are below Sea Level and if a house is put too close to water, and the house has a basement, it will flood . . . So I told her to move the house to higher grounds, and away from the water. Problem solved and I am so relieved that I was able to help someone out with a problem. Now this isn't about creating but I did solve a problem. Many ask me simple questions such as, how do you put furniture on an angle, or how do you place furniture out of the grid squares.

TSR and other *Sims* fan communities have many important properties, and we will discuss these in the next chapter. The point here is that this community is able to incorporate, help, mentor, teach, and guide people from all different backgrounds. Everyone has the opportunity eventually not only to be helped but also to help; not only to learn but also to mentor and teach, not only to follow but also to lead. Even experts sometimes learn and sometimes teach. Those who move faster, in some cases because they have had better preparation for learning in this area, just move faster to help others as well. Their good early preparation is put into service for those who have not had such good early preparation.

Today there is much talk about "communities." People interact with each other in many different chat rooms, Internet forums, fan sites, and interest groups. Many do not fit the rather romantic notions of closeness, support, and belonging we tend to associate with word "community" (probably inappropriately, since communities sometimes can be narrow, exclusive, and hostile to outsiders). But *The Sims* communities that people like Tabby Lou and Izazu took part in do reflect some of these romantic notions. This is not true of every site devoted to *The Sims*. Some are much harsher places, and we will look at this difference in the next chapter.

TSR and similar *Sims* communities use encouragement and support, not negativity and criticism, to help people. They are encouraging to newcomers. They honor experts like Tabby Lou and Izazu, but they do not discourage or demean people progressing at different stages and paces. It is, perhaps, tempting to think that this is all because these fan sites have lots of women on them, though of course, many men contribute as well.

It is unlikely that the presence of women is the main cause of these positive community dynamics. What seems to create and sustain these positive dynamics are a variety of other features. Some of these are a mixture of ages, young to old; a focus on a shared passion that trumps other properties of people, without rendering these properties irrelevant; a shared set of norms about helping people and spreading the passion (since helping people is the best way to spread the passion); respect for diverse ways of and paths to learning and mastery; and a commitment to high standards that members believe everyone can meet if the newcomers get help and bring or gain grit.

When asked what her family thought of her *Sims* passion, Izazu said her family "still thinks I am playing." They do not, she said, "understand the difference between creating for sims or playing them." Her departed husband, though, "was very proud of me for what I was able to do." For women like Izazu, creating is playing. It is the way she now plays *The Sims*. For more and more gamers, male and female, playing games leads to modding and designing, and, in turn, design becomes their favored form of play.

When talking about her family, Izazu also said she is now the only one in her family "hooked on *The Sims 2*." This is only true in the traditional sense of the word "family." There are millions of people hooked on *The Sims 2*, and a great many of them are now, thanks to TSR, in Izazu's extended family.

At the time we wrote this chapter, Izazu's designs had been downloaded close to two million times. She is still going strong. In school, without good early preparation for her learning, she would have been in danger of being seen as "behind" her peers. However, like Tabby Lou, Izazu is a rock star. How many children in school without good early preparation for learning would have become stars if they had not been told they were "behind"?

We told Izazu's story because she is a woman, unlike Tabby Lou, who did not appear to have early preparation for learning to be a *Sims* designer. But we also asked earlier, what might designing for *The Sims* be good preparation for itself? What learning trajectories can it lead to? We do not know the full answer to this question, though it is the sort of question we need to begin asking and answering for all sorts of learning people do in and out of school. We do know that Izazu certainly saw *The Sims* as leading to new learning in other areas.

From her *Sims* work, Izazu developed an interest in architecture in her hometown, in magazines and on the Web. She said, "I enjoy looking at all the old and new architecture we have in my town and basically appreciate it now!" She got ideas for landscaping her own home from the landscaping she did in *The Sims*. She intended to take a 3D art course when we last talked to her in 2008. She readily acknowledges that her computer skills have improved considerably. Izazu also now has an interest in photography. She takes pictures and makes them into art for her *Sims* homes and decorates her own home with them as well.

However, she says that "the most important thing that has changed and improved in my real life from creating for the game" is, "my ability to make things using programs that I never even knew existed before. I make pictures for myself and the game and now family members also. And when I see it makes them happy, I am happy and that makes it all worth it. I couldn't of done it without the experience I have had making custom content for a game." Even for a middle-aged woman, designing for *The Sims* spills over into her life in a variety of ways. She is prepared for future learning in a variety of areas because of her work (which is really a type of play) in *The Sims*. Should she find a new passion, there is little doubt she would be off and running again.

EarthGoddess

We want to close this chapter with an excerpt from our interview with a woman named EarthGoddess. When we interviewed her, EarthGoddess was a twenty-eight-year-old mother of three and in college. She was a featured artist on TSR but has since "retired" as a featured artist there and started several sites of her own. Figure 5.1 is a photo of a few of her many Sims creations. We will comment on EarthGoddess's answers, since they are so typical of what we found in our other interviews.

Figure 5.1 Custom Sims houses, designed by EarthGoddess

Note: Originally posted at TSR (http://www.thesimsresource.com).

INTERVIEWER: What prompted you to start making content?

EARTHGODDESS: Wanting to know how it was done, oddly enough. I had seen so much beautiful custom content out there, I couldn't help but wonder how hard it was to do. I love a challenge! Wanting matching furniture for game play was also a major influence on this decision. :)

What started EarthGoddess's passion was her desire to "match furniture," something seemingly no more significant than purple potties or houses for little people on monitors. What made this so motivating for EarthGoddess was clearly her love of challenges.

INTERVIEWER: What were the first kinds of things you created?

EARTHGODDESS: Homes. I loved building homes. From there, wallpapers and floors seemed the next logical step. While building homes, you start to notice a lack in matching Maxis decor. Learning to recolor objects was next on my To Learn list. And lastly, meshing whole new items. It's really a matter of how tolerant a person is with the build/decorating limitations the game presents, I think. Or maybe it's the obsessed gamer. :) With so many fansites and artists out there, it is also a matter of wanting to learn to do it yourself rather than hunting down or requesting what you want from others.

EarthGoddess noticed a lack in the game—matching decor—and decided she wanted to fill it, a challenge that highly motivated her. Like the other women we interviewed, she learned a good deal from the community and yet was highly motivated to do things for herself. We found among the girls and women we studied that *Sims* designers are highly proactive in wanting to do things themselves and direct their own learning, but they are also nourished in this learning by the resources the *Sims* community gives them and that they give back to it. They see no contradiction between self-directed, proactive learning and community support.

INTERVIEWER: What has motivated you to participate in your Simming activities over time?

EARTHGODDESS: There have been many times I've had to take short breaks from *The Sims* due to a drought in creativity or some real life dilemmas. Gamers requesting specific items for their games and beginning artists needing help has kept me involved. I try to spend as much time encouraging other artists as possible. Having a place like TSR to put your work out there is only a part of it. Having people who support you in your endeavors is absolutely fundamental. I enjoy being one of those people. Like real life, people can be hurtful, especially other artists.

A place to display your work and get famous is highly motivational. Even better is positive support. EarthGoddess enjoys being a certain kind of person, namely, the sort of person who gives and gets from the *Sims* community. She has gained an important new identity through *The Sims*.

> INTERVIEWER: How have you found time to fit your Sims play into your life?
>
> EARTHGODDESS: When the kids are sleeping! LOL Some weeks I have less time than others, real life always takes precedence, but once you're a part of this whole big community of people, it also becomes a priority.

People tend to think of games and gaming as trivial, but that is often because they are unaware of the communities that stand behind games and gaming. EarthGoddess's commitment is not to the game, per se, but to the people in the community built around the game.

> INTERVIEWER: Will you continue to create after you retire [as a featured artist on TSR]?
>
> EARTHGODDESS: Absolutely! There are too many ideas floating around upstairs to give it up completely. I have started my own fansite as well, so I contribute to that. Having your own fansite opens a lot of possibilities and opportunities that weren't available before. It definitely presented me with the opportunity to learn yet another aspect of this community and others: web design! :)

Over and over the girls and women we studied—regardless of age—tell us that they want to keep learning, and they do. They push themselves to keep learning new things. They challenge their taken-for-granted mastery over and over again. This has been said to be the hallmark of real experts (Bereiter, 2002; Bereiter & Scardamalia, 1993).

> INTERVIEWER: What have been the most valuable resources in helping you learn to create content?
>
> EARTHGODDESS: Trial and error. I like to take things apart and see how they work. If it weren't for some of the tools available, such as SimPE, this would be very difficult to accomplish. SimPE is invaluable. Having access to patient people who have been there and done that (and are generous enough to share what they know with me) has also been instrumental. I think there's always a time when you get hung up and need to ask someone with the experience.

Again we see the combination of self-directed, proactive learning and support from the community, and one crucial thing the community has to offer is good tools. Note, too, the importance of having more-experienced and less-experienced people in the same community, even for people who are, in some respects, already experts. You need all different degrees and types of expertise and experience in the community.

INTERVIEWER: What software tools do you use?

EARTHGODDESS Milkshape, UVMapper Pro, SimPE, PSP X2, Bodyshop, Homecrafter, Dreamweaver . . .

Typical of the women we interviewed, EarthGoddess cannot even remember all the software she has mastered—her list trails off in three dots. And these are very demanding pieces of software. These women are all "techies" now, though their focus is not on being a techie but on designing. They master the tools because they need to for their goals and passions, not to "learn technology."

INTERVIEWER: What has been the hardest thing for you to learn? How did you overcome this challenge?

EARTHGODDESS: That my learning is never done. Just when I think I've mastered something, I'm surprised by a new challenge. I think the surprises are actually part of what I love. Overcoming it is simply a matter of having the will to proceed.

Again, learning is seen as never ending. Challenges are good, but effort is required. What EarthGoddess calls "will," we have called grit.

INTERVIEWER: What IT [information technology] skills have you learned from creating Sims content?

EARTHGODDESS: Where to start? There is more than 7 years of IT education to cover. I've learned how to use (and expertly) more than two dozen software programs, something I may never have done otherwise. Being in the community, running into tech issues with game play, helps one get a handle on the basic operations of your computer too. You'd be surprised the things you can pick up on the forums and in the community: hardware such as graphics cards, memory cards, and motherboards (which ones are compatible with other hardware and *The Sims*? which are considered "good"?) How to access and edit screen resolutions and other pc [*sic*] functions. How to properly care for your computer to increase performance: defragging, clean up, virus scans.

EarthGoddess refers to her seven years as a *Sims* designer as her IT education. But it is also excellent preparation for future learning in a variety of technical areas. It is also excellent preparation for future learning in art. For *Sims* designers, art and technology are not in tension, but they are fundamentally married—each leads to more learning about the other.

> INTERVIEWER: Have you used what you have learned from making Sims content in any other area of your life?
>
> EARTHGODDESS: More than I can even begin to describe while not writing a novel. :) Two examples off the top of my head . . .
>
>> My 9 year old daughter is also a *Sims* fan. She loves to design her own furnishings on paper and learn from me how to apply those designs in a computer program—and is sometimes even able to use her designs for school projects. When you play *Sims*, everyone is an interior decorator. :) We *always* enjoy the time spent together. She's learned quite a bit about computers in the process and has learned the virtue of patience.
>>
>> Friendships: I have made many wonderful friends from around the world. They're a huge source of inspiration, support, knowledge, encouragement, and good times. We keep each other striving to do bigger and better, to push our limits—creatively and in life's everyday experiences.

Enough said. EarthGoddess has said it all.

While EarthGoddess is no longer active as a Featured Artist on TSR, she has by no means abandoned *The Sims*. She retired from TSR to create her own *Sims* site, Natural Sims (http://www.naturalsims.com). However, she still has content on TSR. When we last looked at the end of 2009, she had more than two thousand creations on the site and more than three million downloads.

Chapter 6

Passionate Affinity Groups

A New Form of Community that Works to Make People Smarter

Passionate Affinity Groups

Today, success requires passion plus persistence, what we have called grit (Duckworth et al., 2007). We argued in the last chapter that passion often starts small. There is something a person wants badly to do. As we have seen this may be as small as wanting to make a purple potty for a granddaughter or make homes for "little people" on one's monitor screen. The person then finds tools that will let them do what they want to do. Today, these tools are often found within a community of people on the Internet. The person initially goes to the community to get the tools as well as help for using them but then becomes hooked on the community itself and its larger passion; in the case of the women we discussed in the last chapter, this passion is building and designing for *The Sims*.

Everyone we discussed in Chapter 5 talked about the importance of *The Sims* community to their learning, their design work, and their lives. However, there are lots of different types of communities on the Internet and out in the real world (Barton & Tusting, 2005; Hellekson & Busee, 2006; Rheingold, 2000; Shirky, 2008; Wenger, 1998; Wenger, McDermott, & Snyder, 2002). Some are good and nurturing, and some are not. Communities can give people a sense of belonging, but they can also give people a sense of "us" (the insiders) against "them" (the outsiders). People can be cooperative within communities, but they can also compete fiercely for status.

The Sims communities we discussed in the last chapter, communities built around a passion for building and designing for *The Sims*, are communities of a distinctive type. They behave in certain ways that we believe are good for learning and human growth. We have elsewhere referred to communities like *The Sims* design communities as "affinity groups" (or "affinity spaces") or "passion communities" because they are held together by an affinity or passion people share (Gee, 2004, 2007). Here we will call them "passionate affinity groups."

We will soon list a variety of features that a passionate affinity group must have. It is convenient sometimes to invent a new term so we can make it mean just what we want it to mean. We will be able to define passionate affinity groups in such a way that whatever term you want to use (e.g., "community") will not name a passionate affinity group unless it has at least most of the features we list later. There are, of course, lots of other passionate affinity groups besides *The Sims* communities we discussed in the last chapter, but these *Sims* communities are excellent examples of passionate affinity groups. We will discuss later how a community that is a passionate affinity group can cease to be one by losing one or more of the features that define such a group.

We want to argue that human learning becomes deep, and often life changing, when it is connected to a passionate affinity group because of the features that constitute a passionate affinity group. In this chapter, we will identify and describe these features. The following list is the set of features we have found associated with *The Sims* design communities we have studied. This is an "ideal" list; many real communities tend more or less toward these features, thus coming closer or not to being an "ideal" passionate affinity group. For us humans, it is hard to remain close to any ideal, and communities that are close to the ideal can change over time for the worse. However, during the time we studied them, *The Sims* designer communities were very close to this ideal.

As we list the features of a passionate affinity group, it is interesting to think how very different school is from a passionate affinity group. If human learning and growth flourish in a passionate affinity group, then it is of some concern that school as a community has so few features of such a group. To make this point, for each feature below, we will discuss how school compares on this feature. We will, to make the contrast clear, talk about traditional schools or school as we traditionally conceive it. Of course, in this age of school reform, there are many people trying to break the mold of traditional schooling. Nonetheless, it is interesting to see that school is almost the polar opposite of a passionate affinity group.

Features of a Passionate Affinity Group

A common passion-fueled endeavor—not race, class, gender, or disability—is primary. In a passionate affinity group, people relate to each other primarily in terms of common interests, endeavors, goals, or practices—defined around their shared passion—and not primarily in terms of race, gender, age, disability, or social class. These latter variables are backgrounded, though they can be used (or not) strategically by individuals if and when they choose to use them for their own purposes. This feature is particularly enabled and enhanced in virtual passionate affinity groups (Internet communities) because people can enter these spaces with an identity and name of their own choosing. They can make up any name they like and give any information (fictional or not) about themselves they wish. This identity need not, and usually does not, foreground the person's race, gender, age, disability, or social class.

There is an interesting paradox here: what people have a passionate affinity for in a passionate affinity group is not first and foremost, at least initially, the other people in the group but the passionate endeavor or interest around which the group is organized. This is one necessary feature among several others (it is not sufficient) that leads to good, respectful, and encouraging behavior, not hostility and flames. While people in the passionate affinity group may well eventually come to value their fellow members as one of the primary reasons for being in the group, the fact that a shared passion is foregrounded as the reason for being there leads to good behavior because everyone sees that spreading this passion, and thus ensuring the survival and flourishing of the passion and the passionate affinity group, requires accommodating new members and encouraging committed members.

School: Children in school rarely share a common passionate endeavor. In fact, children often have quite different views from each other and from the teacher as to why they are doing what they are doing in school (Willingham, 2009). Too often factors like race, gender, social class, or disability play too prominent a role in school without the choice of the student as how to define and use his or her own identity.

These groups are not segregated by age. Passionate affinity groups involve people of all different ages. Teenage girls and older women, and everyone else in between, interact on *The Sims* sites we discussed in the last chapter (sometimes these are mothers, daughters, and even grandmothers). There is no assumption that younger people cannot know more than older people or have things to teach older people. Older people can be beginners; indeed, anyone can begin at any time. Older and younger people judge others by their passion, desire to learn, and growing skills, and not by their age. At the same time, the older members set a standard of cordial, respectful, and professional behavior that the young readily follow.

School: School is, by and large, segregated by age with only one adult around (the teacher).

Newbies, masters, and everyone else share common space. Passionate affinity groups do not segregate newcomers ("newbies") from masters. The whole continuum of people from the new to the experienced, from the unskilled to the highly skilled, from the slightly interested to the addicted, and everything in between, is accommodated in the same space. Different people can get different things out of the group, based on their own choices, purposes, and identities. They can mingle with others as they wish, learning from them when and where they choose (even "lurking," or viewing but not contributing) on advanced forums where they may be too unskilled to do anything but listen in on the experts). Note that while passion defines a passionate affinity group, not everyone in the group needs to be passionate or fully committed. They must, however, respect the passion that organizes the group; the group will offer them the opportunity, should they wish to take it, to become passionate.

School: School segregates newcomers from more expert students through tracking and grade levels.

Everyone can, if they wish, produce and not just consume. People who frequent a *Sims* passionate affinity group often go there to consume, that is, to get stuff other people have made, and that is fine. But the group is organized to allow and encourage anyone who wants to learn to build and design themselves. Tools, tutorials, and mentorship are widely offered. In other game-related passionate affinity groups, fans create new maps, new scenarios for single-player and multiplayer games, adjust or redesign the technical aspects of a game, create new artwork, and design tutorials for other players. In a passionate affinity group, people are encouraged (but not forced) to produce and not just to consume; to participate and not just to be a spectator.

School: School stresses consuming what the teacher and textbook says and what other people have done and thought. When students produce (e.g., a writing assignment), they do what they are told because they are told to, not what they want because they have chosen it.

Content is transformed by interaction. The content available in a passionate affinity group (e.g., all the *Sims* houses, rooms, furniture, clothes, challenges, and tutorials) are transformed continuously through people's social interactions. They are not fixed. People comment and negotiate over content and, indeed, over standards, norms, and values. Most of what you find in a passionate affinity group, even if you are just there to consume, is a product of not just the designer (and certainly not just the company, e.g., the makers of *The Sims*), but of ongoing social interaction in the group. Content producers in a passionate affinity group are sensitive to the views, values, and interactions of other members of the group.

School: School content is fixed by teachers, curricula, and textbooks, and the students' interactions with each other and with the teacher rarely change anything in any serious way (with the proviso that some teachers, of course, try to adapt material for different sorts of learners, though often without these learners having much say in the matter or as much say as the assessments they have been given).

Both specialist and broad, general knowledge are encouraged, and specialist knowledge is pooled. Passionate affinity groups encourage and enable people to gain and spread both specialist knowledge and broad, general knowledge. People can readily develop and display specialized knowledge in one or more areas, for example, learning how to make meshes in *The Sims* or tweak a game's artificial intelligence (AI) in another game-related passionate affinity group. At the same time, the group encourages and enables people to gain a good deal of broader, less-specialized knowledge about many aspects of the shared passion, which they share with a great many others in the group. This creates people who share lots of knowledge and common ground but who each have something special to offer. It also means experts are never cut off from the wider community. It is important, too, that each person with specialist knowledge sees that knowledge as partial and in need of supplementation by other people's different specialist knowledge for accomplishing larger goals and sustaining the passionate affinity group. Knowledge pooling is enhanced by the fact that everyone in the group shares a good deal of knowledge about *The Sims* and design.

School: In school, most children rarely get to become experts or specialists in anything. Further, the children in a classroom or school rarely share a lot of general knowledge about something about which they all deeply care, which lays the foundation for each child's development of different forms of specialist knowledge, that they can use to achieve common goals.

Both individual and distributed knowledge are encouraged. A passionate affinity group encourages and enables people to gain both individual knowledge (stored in their heads) and to learn to use and contribute to distributed knowledge (Brown, Collins, & Dugid, 1989; Hutchins, 1995). Distributed knowledge is knowledge that exists in other people, material on the site (or links to other sites), or in mediating devices such as various tools, artifacts, and technologies to which people can connect or "network" their own individual knowledge. Such distributed knowledge allows people to know and do more than they could on their own. People are encouraged and enabled to act with others and with various mediating devices (e.g., tutorials and software like Body Shop and Photoshop in *The Sims*, level editors, routines for tweaking the AI of a game, and strategy guides in other game-related passionate affinity groups) in such a way that their partial knowledge and skills become part of a bigger and smarter network of people, information, and mediating devices and tools.

School: In school, individual knowledge is predominant, and distributed knowledge is given short shrift. We are still fighting over whether students should use calculators in math class. There are few deep knowledge tools and technologies around in schools (computers are still too often one to a classroom). Further, students rarely get to trade on each other's knowledge to supplement their own—in school that is often called "cheating."

Dispersed knowledge is encouraged. A passionate affinity group encourages and enables people to use dispersed knowledge: knowledge that is not actually on the site itself but at other sites or in other spaces. For example, in *Sims* passionate affinity groups, there are many software tools available on sites made by the designers of *The Sims*, but there are links to all sorts of other groups, software, and sites that have tools to facilitate building and designing for *The Sims*. In a passionate affinity group devoted to the game *Age of Mythology*, to take another example, people are linked to sites where they can learn about mythology in general, including mythological facts and systems that go well beyond *Age of Mythology* as a game. When a space utilizes dispersed knowledge, it means that its distributed knowledge exists in a quite wide and extensive network. When knowledge is dispersed, strict boundaries are not set around the places from which people will draw knowledge and skills.

School: In school, too often all the knowledge is in the classroom, and students are not linked to sources outside the classroom. In fact, many such links are banned or heavily policed.[1]

Tacit knowledge is used and honored; explicit knowledge is encouraged. A passionate affinity group encourages, enables, and honors tacit knowledge: knowledge members have built up in practice, but may not be able to explicate fully in words. Designers in *Sims* passionate affinity groups very often learn by trial and error, not by memorizing tutorials and manuals. They have their own craft knowledge and tricks of the trade. Players often pass on this tacit knowledge through joint action when they interact with others via playing the game or helping others design for the game. Not all tacit knowledge can be put into words. At the same time, the passionate affinity group offers ample opportunities for people to learn to articulate their tacit knowledge in words (e.g., when they contribute to a forum thread or engage in group discussion about a shared problem).

School: In school, unlike in many workplaces, tacit knowledge counts for nothing. Indeed, students often learn to articulate knowledge (say it or write it down) that they cannot apply in practice to solve problems.

There are many different forms and routes to participation. People can participate in a passionate affinity group in many different ways and at many different levels. People can participate peripherally in some respects and centrally in others; patterns can change from day to day or across larger stretches of time.

School: In school, by and large, everyone is expected to participate in the same way and do all the same things.

There are lots of different routes to status. A passionate affinity group allows people to achieve status, if they want it (and they may not), in many different ways. Different people can be good at different things or gain repute in a number of different ways.

School: In school, there are different routes to status (e.g., being a good student, a good athlete, and other such things). Unfortunately, in the classroom as a community, too often there is only one route to status, that is, being a "good student," which means being good at being a student, not necessarily being good at solving problems or innovating.

Leadership is porous and leaders are resources. Passionate affinity groups do not have "bosses." They do have various sorts of leaders, though the boundary between leader and follower is often porous, since members sometimes lead and sometimes follow. Leaders in a passionate affinity group, when they are leading, are designers, mentors, resourcers, and enablers of other people's participation and learning. They do not and cannot order people around or create rigid, unchanging, and impregnable hierarchies.

School: In school, teachers are leaders and bosses—and often see their role as telling, rather than resourcing learners' learning and creativity—and students are followers.

Roles are reciprocal. In a passionate affinity group, people sometimes lead, sometimes follow, sometimes mentor, sometimes get mentored, sometimes teach, sometimes learn, sometimes ask questions, sometimes answer them, sometimes encourage, and sometimes get encouraged. Even the highest experts view themselves as always having more to learn, as members of a common endeavor, and not in it only for themselves. They want others to become experts, too.

School: In school, roles are not reciprocal. Teachers teach, mentor, and lead, while students "learn," get mentored, and follow.

A view of learning that is individually proactive, but does not exclude help, is encouraged. Passionate affinity groups encourage a view of learning where the individual is proactive, self-propelled, engaged with trial and error, and where failure is seen as a path to success. This individual view of learning is not seen as excluding asking for help, but help from the community is never seen as replacing a person's own responsibility for his or her own learning.

School: Ironically, in school, getting help often counts as "cheating," and yet few students take on a proactive, self-propelled, and engaged trial and error approach to learning. In a passionate affinity group, getting help of the right sort often does lead to such a proactive view of learning.

People get encouragement from an audience and feedback from peers, though everyone plays both roles at different times. The norm of a passionate affinity group is to be supportive and to offer encouragement when someone produces something. This support and encouragement comes from one's "audience," from all the people who have responded to one's production. Indeed, having an audience, let alone a supportive one, is encouraging to most producers. However, producers get feedback and help (usually also offered in a supportive way) from other producers whom they consider either their peers or people whom they aspire to be like some day. At the same time, who counts as a peer changes as one changes and learns new things. Everyone in the passionate affinity group is audience for some people and potential peers for others.

School: In school, children rarely have any audience who really cares other than the teacher, and feedback comes, by and large, from the teacher, who is not a peer (not simply in the sense of age, but also in the sense of expertise) or someone most students aspire to be like, in terms of what they have a passion for producing and learning. School is not a source of encouragement for many students.

The list above is based on the *Sims* design groups we have studied. Other groups have these features as well. It is possible to implement these features in face-to-face communities as well, but it can be harder due to institutional constraints, preexisting status differentials, and even geographical boundaries that prevent people with common interests from coming together. They are not easy to achieve, and they can deteriorate over time. Communities with these positive learning and growth features are miracles of human interaction. We need to know a great deal more about how they start and can be sustained. We also need to study how they can be implemented in areas we as educators and national and global citizens care about.

School, Content, and Knowledge

Many people will say that the contrast between passionate affinity groups and traditional schooling is unfair. They will say that people choose to be in passionate affinity groups. Schools have to force people to do things they do not want to do. In passionate affinity groups, people share a passion. Schools cannot be about passions, since everyone has to do, learn, and know the same things, namely, "what every educated person ought to know" (Hirsch, 1987). Too often, this leads to everyone knowing next to nothing, or at least nothing very deeply.

Here is the sad fact: Humans do not learn anything deeply by force. Humans do not learn anything deep without passion and persistence (i.e., grit). That is just the way humans are made. That, too, is why, for most people, what they learn in school is highly transitory unless they practice it in work or other settings after school. It is also why so many people, children and adults, learn more important things in their lives out of school than in it.

Think, for example, about learning geometry. Forcing people to learn geometry all in the same way because they are "supposed" to know geometry is not effective. Few people learn it well enough to remember and use it unless their jobs give them practice with it after school. They take geometry (or chemistry or algebra) to get to the next level of schooling. It is a gateway. Some people master it at school because they choose to and have a passion for it, if only for a high grade and getting into a good college.

Now consider how geometry learning happens when someone wants to design things in *Second Life*. The building tools in *Second Life* are software tools for designing three-dimensional environments. They require a big learning curve to master, a learning curve that people take on by their own choice, driven by their passion for being a designer in *Second Life*. These tools require one to use a good deal of geometry to get all the angles and shapes to fit perfectly together. In fact, the tools build in some nice representations of geometrical information.

We met a woman, whose story we will hear in Chapter 8, who is a skilled designer in *Second Life* and widely respected. She failed to learn geometry well in school but now feels confident in her geometrical knowledge. This woman did not learn geometry because someone told her she "had to" or "should." She learned it because she wanted to design in *Second Life*, and geometry is required to do that. Further, she had the support of the passionate affinity group in *Second Life* devoted to design. Geometry was a tool she needed for something she wanted to do.

The things we teach in school, subjects labeled "algebra," "physics," "civics," and so forth, are all tools for doing things (Gee, 2007). For example, "physics" is a set of tools for doing physics, that is, for solving problems in physics. These tools can also be used in other enterprises, for example in building roller coasters in *RollerCoaster Tycoon* or designing rockets in real life, much like geometry is used in designing for *Second Life*. "Civics" is a set of tools for being able to understand and participate in government and society. These tools, too, can be used in other enterprises, for example, in designing virtual worlds with their own economies and governing structures. Humans learn things like facts, information, and principles ("content") well and deeply only when they are learned as tools for doing things that are meaningful and important to them (diSessa, 2000; Gee, 2004; Shaffer, 2007).

This brings us to "knowledge," or what school is supposed to be "about." Lots of the features of a passionate affinity group listed above use the word "knowledge." Indeed, passionate affinity groups are, in a sense, knowledge communities. Such groups build, transmit, sustain, and transform knowledge. But this knowledge is always in the service of something beyond itself. This does not mean such knowledge has to be practical in the sense of serving the needs of society as a whole. But it has to be in the service of doing, that is, in the service of solving problems.

In a passionate affinity group, people do not judge what other people know by asking them to list what they know and to write down the facts, information, and principles they know (i.e., what they have stored in their heads). They judge what other people know based on what they can do and how they can put their knowledge to work in solving problems for themselves and in helping others to solve problems.

The philosopher Wittgenstein (2001, p. 52) once said that we know whether someone knows something if they know "how to go on" in a course of action. If someone is doing something, they have to act. Then they have to ask themselves, did my action work and did it bring me closer to my goal or not? If the answer is "no," then they have to choose how to "go on," or how to proceed on a trajectory of actions that will, eventually, lead to success. All the knowledge in the world will do you no good in geometry, civics, or designing for *The Sims* if you do not know how to assess the success of your actions and how to go on in a successful trajectory to accomplish your goals (and sometimes one way to go on is to change your goals). This is the main thing passionate affinity groups teach.

The learning scientist Dan Schwartz at Stanford University has said that looking at the choices people make in a course of actions devoted to solving problems in a certain area is a much better assessment both of what they know and of how well prepared they are for future learning in the same area (personal communication, see also Schwartz, Sears, & Chang, 2007). He suggests we should teach and assess choices, not knowledge, as content. For example, in solving a problem in science or mathematics, or of designing a building in *Second Life*, what are good choices to make when something has not worked? Should one try multiple solutions even if one solution already works? Is it good to write down representations on a piece of paper as one goes along or leave everything in one's mind? Imagine the transformation in schools if learning in school became about how to make good choices in science, mathematics, art, and civic participation.

Passionate affinity groups are organized to help people make better choices. They are organized to share information so that new and better choices can be discovered. They are organized, as well, to share information about choices that work and ways to learn how to make better and better choices. These choices are not just about designing things. They are also about how to socially interact in the passionate affinity group, and outside it as well, including in "real life," so that goals are accomplished and people grow, no matter what their age. When this focus on discovering and making good choices lessens, passionate affinity groups deteriorate. They become sites of mere socialization and fights arise over status, belonging, and how to behave.

Bell Curves and the 80/20 Principle

The ways in which participation and production work in a passionate affinity group are quite different from school. Schools operate by the bell curve. In a bell curve, the great majority of people are in the middle range of achievement, with a few much better than the rest and a few much worse. Passionate affinity groups, and other interest driven groups like *Flicker* (a photo sharing site), for example, tend to operate by the principle we mentioned earlier, the "80/20" or Pareto principle (Shirky, 2008). Eighty percent of the people in a passionate affinity group produce 20 percent of the content (the designs, pictures, or whatever the activity of the group is) and 20 percent of the people produce 80 percent of content.

This 80/20 organization means such groups can recruit everyone's contributions while allowing the most dedicated to produce a great deal more. If we believe that young people today learn a great deal in such interest-driven groups, then it is important that there are lots of them and that everyone can find ones in which they can be in the 20 percent of high contributors, if they wish, while making contributions in others where they are in the 80 percent.

Many people think that the bell curves we find in school, where nearly everyone is clumped in the middle at average, are just a reflection of people's "natures," that is, their genetics, like a normal distribution of height. Most of us, we think, are average performers and only a few are really good or really bad. But in reality, as Gould (1981) long ago pointed out, the standardized-testing industry assumes bell curves and designs tests to get them. The design and scoring method of such tests is normative, just as is grading "on a curve." There must always be some students who have lower scores than all others, and some who have higher scores, even if the actual difference in their performance is quite small. In addition to how tests are designed, the way that schools design instruction also contributes to an artificial view of people's abilities to learn. When people are organized to learn something like algebra, with little choice, passion, or lucid understanding of why they are learning what they are learning, the result is a bell curve. Most people cooperate and learn something, if not much. A few resist and learn nothing, and few find their own deep reasons for learning algebra. It is not that some people simply are not "gifted" at something like mathematics. What people learn outside school shows that nearly anyone can learn such things if they need and want to do it. Consider the woman we discussed previously, who hated geometry in school and now uses geometry regularly, with confidence, because she has a passion for building in *Second Life*, and such building requires geometry.

Not everyone has a passion for the same things. People join groups that support their learning and resource them. In some cases, this is enough. In some cases, they get hooked on the community and the passion the community supports and join the top 20 percent.

Jade, the young girl we discussed in an earlier chapter, designed for *The Sims* and uploaded her designs to *Sims* sites. These communities resourced her and helped her to become an expert designer. However, she found her passion not in *Sims* communities but in *Second Life*, where she "transferred" her skills from *The Sims* to building and designing in *Second Life*.

For Jade, *The Sims* was a way station on her trajectory of learning and growth as a proactive learner and designer. We badly need to study and understand such trajectories before we judge anyone's participation in a passionate affinity group or any other sort of learning space. *The Sims* was a deep preparation for future learning and participation for Jade. That she was in the 80 percent of *Sims* designers was no indicator of lack but rather of her stage of development.

Creating Passionate Affinity Groups

There are today innumerable interest-driven groups on the Internet. Most of these are not passionate affinity groups. They are not passionate affinity groups unless they have the features we listed above, and these are not easy features to design and sustain. Today, many businesses start their own interest-driven groups on the Internet, companies such the Saturn car company, Weber Grills (which has a site called *Weber Nation*), Diesel jeans, Coca-Cola, and Kleenex.

Most company-sponsored, interest-driven groups do not come close to being passionate affinity groups, though the sites often claim that people should and do have a passion for their brands. Such sites let people engage with each other socially, play games, and often build, design, or otherwise produce something relevant to the brand. These companies want people to see their brands not as products but as experiences that people can share as part of their identities and lifestyles (Wasik, 2009).

Of course, any company that had—as does *The Sims*, which is, after all, connected to a big business (Maxis, which is owned by EA, a very big business indeed)—a real passionate affinity space would think it had struck gold. After all *The Sims* as a brand has millions of people working for it by designing game content and identifying with it.

Anyone, however, who has ever heard Will Wright, the creator of *The Sims*, talk about his games knows that he cares more about resourcing people's creativity than making a profit, though we are sure he sees the two as intimately connected. He knows well that, to resource people's creativity, you cannot try to dupe them. You have to give them real and powerful tools, trust in their creative capacities, listen to them, and grant them a good deal of freedom, including the freedom to criticize *The Sims* and the company that makes it.

Wright has often changed features in his games because of what he saw his players do and say: he has often let them lead him. But Wright is making a product whose whole point is to resource people's creativity on the assumption that everyone who wants to can be creative and innovative.

Passionate affinity groups grow in the soil of good tools and good social interactions. They cannot be directly constructed and made to follow the dictates of a company or boss. People who want to make such groups better see themselves more as gardeners than as bosses, and gardeners of a garden that will, in many ways, go its own way, finding deep roots and nourishment and, indeed, new plants, where and when the gardener did not anticipate.

At root, a passionate affinity group has to be built around something worth having a passion about. That is part of Will Wright's genius. He saw that making houses for little people on a monitor, or even purples potties for granddaughters, were things that could lead to a passion that was worth having and nourishing. He also saw that this passion could unleash new passions and start people on deep trajectories of learning.

Tools and the Great Passionate-Affinity-Group Paradox

Good tools are foundational to passionate affinity groups. In such groups, people produce and do not just consume. We have argued that people often find their way to such groups because there is something they need and badly want to do. The group has a tool that will let them do it, and it will help with learning the tool. In the end, often the person becomes hooked on the community and its larger passion.

The tools that channel passion, that lead from initial desire to full membership in the passionate affinity group, are at the heart of a deep paradox in passionate affinity spaces. The tools, like Adobe Photoshop and the many other sorts of software that *Sims* designers use, are challenging. Since they are challenging, people need to get help and mentoring. They must and often eventually join the community that gives them this help and mentoring. As they master the tools, they become experts in the community, offering help and mentoring to others. They have status and respect in the community.

However, companies like Will Wright's Maxis want to make creativity available to more people. They want to get more people involved, excited, and producing and not just consuming. They want to do this because the more people who create, the more profit the company makes when people are creating around the company's products (which includes *The Sims* and their other games). But companies like Will Wright's most certainly want to do it because, like any good educator, Will Wright believes in people's capacities and wants to resource as many people as possible to be creative.

Thus, there is an impetus to keep making the tool more user friendly, less challenging, more inviting for more people. Wright's recent game, *Spore*, is about developing from a single-cell creature through every aspect of life on Earth, to cities and civilization, and eventually out into the stars to terraform planets with new life. In *Spore*, Wright offers incredible design tools with which players can design fantasy creatures, buildings, vehicles, and whole planets. The game is made in such a way that you cannot even play it without building. It is a game where you play, then build, then play, and then build again, and so on until the end.

The building tools in *Spore* are so user friendly that more people can engage in design. Indeed, every player must. As you might expect, *Spore* has unleashed a torrent of creativity, and, indeed, passionate affinity groups have evolved that carry this creativity to ever higher and more passionate heights.

New, more user-friendly tools open up creative opportunities to more people. But such tools also, in a sense, put people who are already experts with the old tools "out of business." More people can do what the old experts can do. The designer who struggled to master a difficult building tool or piece of software, and earned status by helping others to learn it, now watches others create great things without such a challenge, as they are now given much more user-friendly tools.

It is the challenge of a good tool, a tool someone needs and wants, that often brings people to the community in the first place. They need help and mentoring with the tool and other related tools they will discover. If the tool is too friendly, they need the help and mentoring of the community less. Of course, they still need the community for an audience and for feedback, as well as for socialization around their shared passion.

Tools and communities will always be in a cycle, where new, more user-friendly tools open up more creativity to more people, but lessen the meaning of people's earlier struggles and achievements as well as the need for more intensive help and mentoring, the point at which many people get hooked on the community and its larger passion. We do not want to stop this cycle: giving more people access to higher levels of creativity is a wonderful goal. But we also have to remember that challenge, when placed inside a nourishing community, is itself good, as a hook to belonging and moving ever further on a trajectory of learning.

There is also the issue of how skills in one area are good preparation for learning for in other areas. When tools are more challenging, and when many of them (like Adobe Photoshop) do not come from the community itself and are not single-purpose, user-friendly versions just for the purposes of the community, they serve as good preparation for future learning in other areas. They can connect learners to other areas and passions and can lead to multiple trajectories that lead to multiple places. A more single-purpose tool (say a tool for recoloring objects that works well and in friendly fashion but only for one game or one community's purposes) cannot carry learners out into a world of other passions and possibilities.

This is a dilemma, indeed: how do passionate affinity groups let more people in but still make it worthwhile to be in? The only moral we have to offer here is this: do not forget that to humans a good challenge, with the right help and mentoring, is a profound, life-changing learning device.

A Harsh Community

In the last chapter, we looked at women participating in two sites that constitute passionate affinity groups: *The Sims Resource* (TSR) and Mod The Sims 2 (MTS2). These sites offer a good deal of support and encouragement for people with quite diverse skills and backgrounds in *The Sims*. Not all sites devoted to *The Sims* operate like these sites.

An interesting contrast to these two sites is a site called *More Awesome than You* (MATY). This is a site whose participants pride themselves on being at the "cutting edge" of *Sims* hacks and mods. The participants are, for the most part, quite technically adept. The norms of behavior for the site favor dealing harshly with anyone with whom one disagrees and especially with newcomers ("newbies") or people who are not highly skilled.

MATY is not, using the phrase of one post, "a standard buddy-buddy forum" (Rohina, 2009, reply 25). The post goes on to say that "you don't come here to be loved or fawned upon or greeted with open arms, you come here for information and downloads to make your game More Awesome . . . If you don't want chunks bitten off you, don't play with tigers." Another post includes the admission that MATY regulars "tend to give new arrivals a particularly hard time." The post says that this "affords us a great deal of entertainment." It is also a way, the post says, to separate the "wheat from the chaff" and keep only newcomers who are tough and skilled (Rohina, 2009, reply 94).

MATY members are no fans of sites like TSR. In fact, MATY contains one thread that is a vicious rant against TSR site owner, whom the thread accuses of "brown-nosing" Maxis in order to get prerelease access to *The Sims 3* and trying to co-opt *Sims 3* modding tools (Merlin, 2009). In any case, to MATY members, TSR would probably be a "standard buddy-buddy forum," and, in the words of another post (which tells people who do not like MATY that they can go elsewhere), "There will be other places where you can have a group sing-a-long of 'Kumbaya' and pretend to care about each other's days, your Special Sisters, your 'creative' abuse of the English language, your made-up attention-seeking disorders and diseases, and your emotional ups and downs" (Rohina, 2009, reply 43).

MATY has many features we associate with a passionate affinity group. It is clearly a site of very high-level knowledge production—its mods are among the best available. Indeed, MATY had out an extensive mod of *The Sims 3*, a mod that corrected many errors in the code and made many improvements to the game, within two weeks of its release. However, its failure to accommodate a wide diversity of skills and backgrounds, and its treatment of newcomers, make it by our definition not a passionate affinity group or, at least, only a partial one.

Designers like Tabby Lou are not respected on a site like MATY. In fact, the site contains several criticisms of Tabby Lou. MATY participants look at themselves as hard-core technical experts. Tabby Lou and many of the other women we studied do not view themselves as such hard-core experts.

They appear, rather, to view themselves as advanced learners. Furthermore, they see their expertise as part of the community, something that adds to the community but is also always supplemented by the community. Finally, women like Tabby Lou do not see their technical and design expertise as separate from the social relations they have contracted in the community and the emotional intelligence they seek to combine with their technical expertise.

How people behave in these communities is not, in fact, a fixed property of them as individuals. It is certainly not due just to the presence of women or men. There are women on MATY and men on TSR. In fact, we have tracked the same individuals engaged with both sites. On MATY they behave harshly, and on TSR they behave cordially. How these communities behave is ultimately a matter of the culture a group grows and attempts to sustain.

We do not have a label for "experts" like Tabby Lou, though we are much more familiar with the sort of hard-edge, high-tech expertise of many of the MATY participants. However, we live now in a world where individual expertise, especially expertise that overvalues what it knows and undervalues what it does not know, is dangerous, as was the case with Alan Greenspan's inability to predict the current global economic meltdown, as mentioned previously (Andrews, 2008).

Things are too complex today to trust individual experts. As we have said, they tend to trust their knowledge too much and pay too little attention to what they do not know, and to what others, perhaps those quite unlike them, do know. We need to grow not expert individuals but knowledge communities. Some will undoubtedly be like MATY, which for all its harshness does network people together into a knowledge-building community. But some will be like the passionate affinity groups we have discussed—communities of shared learning devoted to spreading passion and knowledge and not restricting it to hard core experts. True innovation is as likely, or even more likely, to grow in a diverse community, with lots of different skills and backgrounds, than in a more narrowly defined community, no matter how high the status.

Chapter 7

A Young Girl and Her Vampire Stories

How a Teenager Competes with a Best-Selling Author

Fan Fiction

Fan fiction—fiction written by fans of a game, movie, television show, or book and based on the characters—is a massive enterprise today. For example, there are hundreds of thousands of stories written by fans of the *Harry Potter* series on a variety of Web sites (see, for example, http://www.harrypotterfanfiction.com/ and http://www.fanfiction.net/book/Harry_Potter/). There are a myriad more on other topics written by fans of nearly any media product one could think of, including some long disappeared from mainstream attention. On fan-fiction writing sites, writers gain instruction, mentoring, feedback, and an audience (Black, 2008; Hellekson & Busse, 2006; Jenkins, 2008).

The large amount of fan writing today is coupled with a great deal of other types of writing on the Internet. This other writing is composed, in part, of technical writing such as strategy guides and modding instructions for games and the sorts of design-based discussions we saw in Chapter 5, as well as a great many other technical matters fueled by the massive amount of production carried out by professional amateurs (pro-ams) today. There are also forums on the Internet devoted to an endless array of topics, technical and otherwise, where people share information, seek help, and mentor others.

All this writing is interesting from a historical perspective on literacy. Throughout the history of literacy, including modern times, reading was much more common than writing. In fact, many people could read well but not write well, either because they had never really learned to write or because they rarely practiced it. This is still, in fact, true today, even in developed countries like the United States (Gee, 1990).

Historically, writing was often feared by political leaders, religious leaders, and elites because it was thought that writing might lead people to agitate publicly for their own views against the powers that be (Gee, 1990; Graff, 1979, 1987). Indeed, this has been one effect of communication on the Internet as people have been able to share information about political and military matters—in countries like China and Iran and a great many others, including, of course, the United States—that were easily hidden and denied in earlier times (see, for example, Grossman, 2009; Kennedy, 2007).

Fan-fiction writing and many other types of writing on the Internet have effects other than political agitation. Such writing is a crucial part of the blurring today between "professionals" and "amateurs" (Leadbeater & Miller, 2004). Today, people in every profession—law, medicine, computer science, news, photography, film, games, among others—compete with amateurs who create and share their own knowledge, advice, and products on the Internet, often collaboratively. For example, in the field of fiction, for decades we have thought of fiction writers as professional writers associated with professional editors in official publishing firms. While some decades ago most fiction writers worked in some other job or tried to make their living through their writing, today a great many professional writers work as faculty members teaching writing in colleges and universities.

We will see in this chapter that amateur writers today compete with professional ones for audience and adulation. We will see, as well, that today even some professional writers show deep similarities to the very writers who write fan fiction based on their work.

Fans writing stories based on *The Sims* is a big enterprise. Such *Sims* stories are part of the larger universe of fan fiction. But *Sims* stories are different. They involve pictures and words. In fact, they are a bit like story boards for a movie or game. In their most basic form, *Sims* stories are created by taking screenshots during *Sims* game play. Within *The Sims 2*, the pictures are automatically saved and can be viewed within the game as an "album," while *The Sims 3* requires players to upload screenshots to an online story-creation tool with more elaborate design tools. Players write captions under the screenshots, creating a continuous narrative. Players have developed a wide repertoire of strategies to improve the quality of their pictures. They use cheat codes and custom content to design characters and "sets" for their stories and edit the photos with graphic-design tools such as Adobe Photoshop. Some *Sims* stories are quite elaborate, with hundreds of pages and many chapters.

Players have posted guides for every aspect of story creation, ranging from how to write good dialogue to posing your Sims (which turns out not to be an easy task). *Sims* stories take many forms. Sometimes guides and tutorials are created using the story album, but the vast majority of stories are fiction of some sort. Some stories are based on television characters, movies, or books, but most of the *Sims* stories are about original characters made up by the author.

Sims stories created with the in-game or online album tool can only be uploaded and viewed on the official *Sims* Web sites because special viewers are required. However, there are sites like *The Sims 2 Writer's Hangout* (SWH; http://similik.proboards.com/) where *Sims* writers post links to their chapters on the official *Sims* sites and then set up an individual discussion thread where readers can comment on their stories. They also post individual screen shots from the game as "teasers" for their stories.

At the time we were writing this chapter, there were 161,867 story entries on TheSims2.com, the official fan site created and moderated by Electronic Arts (EA), the company that now owns the *Sims* franchise. Since stories are often posted as individual chapters, one story can have more than one entry. "Romance" is by far and away the most popular genre: 48,754 stories were labeled as romance (stories can have multiple labels). In comparison, 6,513 were labeled "Horror" stories. The release of *The Sims 3* is fueling yet a great many more stories.

Vampire romances are a popular genre among teens writing *Sims* stories. When we searched the stories by the keyword "vampire," 1,051 were identified, and this may underestimate the total, since authors might not have used vampire as a keyword even if they included vampires in the story. The popularity of vampires is due, in part, to the fact that vampires are readily created with a cheat in *The Sims 2* and also, more importantly, because of the prevalence of vampires in popular media such as the *Twilight* series of books, the *Buffy the Vampire Slayer* television series, and the current HBO series, *True Blood*.

Alex and Stephenie

SWH is where we discovered Alex. We were looking for stories that had generated lots of discussion and Alex's certainly had. Her most recent series, "Lincoln Heights," is a vampire romance inspired by the famous *Twilight* novels (and now movies), novels that are incredibly popular among teen and preteen girls. Alex's stories had their own forum thread that had garnered many pages of comments and 15,046 page views in just over a year.

What should one make of Alex? She will raise a good many issues for a good many of our readers. And different readers, we are quite sure, will respond to her quite differently. Though only a teen, she is a popular writer with an adoring audience of other teenage girls. She cannot spell well and her discussion with her fans is carried out in the digital-age teen-speak[1] (replete with typing errors) that so many adults cannot understand. She revels in a world of teen romance with her fans that some adults will see as immature, romantic emotionalism and teenage over-the-top enthusiasms.

Others will see Alex as an icon of our digital world in which more and more people can produce work in competition with established professionals (in this case, Stephenie Meyer, the author of the *Twilight* series). They will see Alex as a rising star, learning to write and getting tech savvy while she gains an audience of fellow teens whose culture she fully shares and collaboratively enacts in a shared search for meaning in their lives.

Alex is, in many ways, a sort of Rorschach test for people's attitudes toward popular culture. Alex is the paradigm of what the media theorist Henry Jenkins (Jenkins et al., 2009) has called "participatory culture": "A participatory culture is a culture with relatively low barriers to artistic expression and civic engagement, strong support for creating and sharing one's creations, and some type of informal mentorship whereby what is known by the most experienced is passed along to novices. A participatory culture is also one in which members believe their contributions matter, and feel some degree of social connection with one another (at the least they care what other people think about what they have created)" (p. 3).

Jenkins argues that such participatory culture is the hallmark of today's popular culture. More and more people want actively to participate in cultural production, not just to watch others who are "stars." Such participation is tied to production. Many young people today (and older ones, as we have seen in Chapter 5) want to produce their own music, video, games, fiction, art, and a good many other things. They want to produce and not just consume. In addition, they want to be "stars" themselves, to have an audience, and not just to be a spectator in someone else's audience. Digital media opens these doors for more and more people.

But before we try to understand Alex, let us turn to Stephenie Meyer, the author of the best selling *Twilight* books, and what she represents. Stephenie is a thirty-six-year-old Mormon mother of three boys. She began writing *Twilight* when she was twenty-nine. On her Web site (http://www.stepheniemeyer.com/), she credits the origin of *Twilight* to a vivid dream, a dream that prompted her to start writing the first novel in the *Twilight* series. She mentions using the Internet for inspiration as well as to get information about publishing (she found WritersMarket.com and Janet Evanovich's Web site, http://www.evanovich.com, to be particularly helpful) and finally got the help of a literary agent online through a group called Writers House. The agent helped her edit the book and send it out to publishers, including the one that ultimately accepted it, Little, Brown and Company. We will see later that Stephenie has a vexed relationship with her professional editors.

Stephenie Meyer is not all that different from Alex, though she is older. She was not a professional writer, and she was inspired to do fiction in a genre, vampire romance, that is, while not new, close to the hearts of teenage girls. Her writing is not notable for its excellence as "art," but her tales are indeed notable for the excitement they have created among girls and women and the money they have made (Grossman, 2008).

It is popular for academics and other "sophisticated readers" to dismiss romance fiction as trivial daydreaming and wish fulfillment. Worse, they say that such stories just reproduce and sustain gender stereotypes and other mainstream cultural assumptions without challenging them. Such stories do not constitute a challenge to taken-for-granted cultural assumptions and the ways in which these assumptions can ignore diversity and marginalize people who not fit standard stereotypes (e.g., girls not into teen romance or who see it as oppressive to their sense of self).[2] But romantic fiction and other popular-culture writing is not unique in this respect. Consider the functions of myth in cultures like Homeric Greece or the Semitic cultures that gave rise to the Old Testament stories. In cultures with bodies of mythic stories, these stories did not exist to challenge people's assumptions about self and society but, rather, to tell who they were then and what their lives meant in some larger sense. Like people today, everyday people in these cultures often felt small and insignificant in the larger scheme of things. Mythic stories told them that they were connected, far in the past, to a set of larger-than-life ancestral figures (e.g., gods, giants, or heroes who once walked the land). Everyday people were encouraged to see their lives as replications of what these gods and heroes had done and as continuing their legacy. This sense of reenactment and legacy gave everyday people's lives a larger meaning and place in the universe. They walked in the footsteps of gods and heroes, and they continued a story that had started long ago and had deep meaning, a story they were continuing, even if in an attenuated or "fallen" form.

Today, while we no longer have robust mythic systems, the world's major religions (once their oral tales became written down) serve, for many people, much the same purpose as myth did. They tell people who they are, why they are here, and where they are going. Religion makes them part of a bigger story and connects them to gods and heroes (the American story of our origins and founding fathers does much the same thing for many Americans).

In the modern world, however, many people want to give meaning to their own lives, to make up their own stories. This trend started in the Enlightenment and has accelerated ever since, most certainly fueled by today's digital media and participatory culture. Today, many people want to tell stories that let them walk in the footsteps of ancestors, gods, and heroes storied not by professional bards, shamans, and priests, but storied by themselves. That is, today many people, including teenagers like Alex, want to minister to themselves and their peers. They themselves want to be "shamans" or "priests" telling stories that give meaning to life and one's emotions as a human being.

When Stephenie Meyer tells a vampire romance, she is not writing as a professional artist, giving people higher meaning sanctioned by high or elite culture. She is, as we will see later, a woman not all that different from the teens who read her, writing out of her own dreams and desires as an everyday person and not as a "professional" who knows more and better than others about what their own lives mean. She is telling a story that others can readily create or modify and tell themselves. For some teen readers, Alex is even better at telling such stories; she is even closer to them, their lives, and their emotions.

The purpose of myth, religion, and the sorts of stories that Stephenie Meyer and Alex tell is not to challenge people to think, reflect, and critique while keeping everything open to constant reconsideration as we learn more and become more tolerant (the purpose of some academic critical writing). People have to get on with their lives in a difficult, dangerous, and, unfair world. They crave meaning, especially meaning that will get them through difficult times and emotions.

Neither Stephenie Meyer nor Alex is producing "critical" writing that challenges society and culture, so, many academics will see it as trivial, ideological, and even dangerous. But there already is lots of critical writing. It is necessary and good. Yet people also need support on their journey in life and cannot always critically reflect on everything as they try to live and cope (e.g., as a teenager longing for acceptance and love). There must be a balance, and the real danger is when this balance is lost. Teenagers, of course, are famous for challenging everything adults accept. We forget this when we watch them consume and now, more than ever, write things like romance fiction, seeking a story that makes sense of their longings and fears. Perhaps they know better than adults that there must be balance between reinforcement and critique.

Here is what one teen has to say about Alex's most famous story, "Lincoln Heights." It is clear that this teen fan found Alex's writing helpful in a deep way: "i LOVELOVELOVE Lincoln Heights. When i read it it always seems to cure my sadness and it has actually helped me deal with alot of depression and shit i've been forced to deal with lately. I can't wait until the next chapter is out . . . HURRY UP, ALEX! xo. :]" (LaurenplayerP, 2009).

Though we ourselves are academics who have studied these matters, we do not think we know Alex's audience and their needs better than she does. We are another set of professionals, like professional writers and animators, with whom she is in competition. As far as we can see, she does very well indeed. She is not the only writer teens ought to read, but she is, indeed, one they should. The girl whose comment we just quoted does not care about good spelling or waiting for critique alone; she is waiting for meaning, meaning of the sort she can eventually make on her own, perhaps in her own stories.

Alex

Alex is on her way to being a star, though she is only a teenager. Her biography written for SWH, "Alexvamp15" (which is Alex's username on SWH, but we will call her "Alex," as do most of her fans) says she was born on July 7, 1993, and that she was fifteen when she joined SWH in January 2008. However, she was active on *Sims* fan sites well before that time. She created a *Sim* page on TheSims2.com in July 2006, right after her thirteenth birthday. She uploaded the first chapter of her first *Sims* story a few days afterward. Here is the description she posted with her first chapter of that story: "Allie's just a normal teenager. But when things go wrong she's so much more than normal."

Alex's first story was a pretty typical teen romance, but it was longer than usual for *Sims* stories, with sixteen chapters uploaded in a period of four months. The chapters initially were shorter than her later creations, but by the last chapter she was up to 101 pages. Alex has devoted huge amounts of time to her stories. Her writing as well as her pictures have improved markedly over time. Indeed, many of her readers comment on this improvement.

Alex never finished her first story. Rather than finish it, she started several new stories, one of which was "Confessions of a Teen Idol/Geek" (Monkey-GalPly, 2007a), and she posted the first chapter in February 2007. Her short description of this story says, "Tanzie's life can't get much worse, but it can get a lot better." This story has fourteen chapters, posted from February to May 2007. With this story, Alex starts a practice of listing popular songs as "themes" to accompany her stories and their characters. Here is the song list for the first chapter of *Confessions* (MonkeyGalPly, 2007a):

> "Story Theme Song: What You Waiting For?" by Gwen Stefani
> "Tanzie: Girl Next Door" by Saving Jane
> "Gabriel: Far Away" by Nickelback
> "Damien: Just The Girl" by The Click Five
> "Maddy: Girl In The Band" by Haylie Duff

The theme songs add another dimension to the visual and written elements of her work and a connection to popular culture familiar to many of her readers. These songs reinforce the romantic themes in "Lincoln Heights" and give further weight to their significance for her audience.

"Lincoln Heights"

The story "Lincoln Heights" (MonkeyGalPly, 2007b), posted chapter by chapter over months, became Alex's big hit. Alex posted the first chapter of "Lincoln Heights" on November 12, 2007, about five months after she posted the conclusion of "Confessions." She started several stories after "Confessions," but "Lincoln Heights" is the story that really earned her an adoring audience.

To date, Alex has posted twenty chapters (two were so long that she split them into two parts) of "Lincoln Heights." The first one was posted on November 12, 2007, and the most recent chapter at the time of this writing was posted in July 2009 on her new Web site (http://lincolnheightsseries .webs.com/29mylastbreathi.htm). The chapters individually are quite long. For example, the last chapter posted on TheSims2.com has 101 pages, which each consist of a screenshot and accompanying text.

As we said above, "Lincoln Heights" is indebted to the *Twilight* book series by Stephenie Meyer, a popular series of books about a girl and her vampire boy friend. Here is what an editorial review on Amazon.com has to say about Meyer's first book in the series: "Bella Swan's move to Forks, a small, perpetually rainy town in Washington, could have been the most boring move she ever made. But once she meets the mysterious and alluring Edward Cullen, Bella's life takes a thrilling and terrifying turn. Up until now, Edward has managed to keep his vampire identity a secret in the small community he lives in, but now nobody is safe, especially Bella, the person Edward holds most dear" (Amazon.com, n.d.).

Let us point out, before we tell more of Alex's story, that this sort of plot, at a deeper level, is not dissimilar in its themes from Jane Austen's work. Austen lived at a time and in a world where women were dependent on the resources and reputation of the men they married. Early capitalism had become powerful enough that many men were rich, but not necessarily well born or raised in the elite culture of the aristocracy. Indeed, there were "men on the make," as there are in our capitalist world. It was harder than in the old days of the aristocracy to know whether a suitor really had money or potential and was ethical and cultivated, or just pretending. Austen's books are full of women thrilled with a new suitor but terrified of what will happen if things go wrong or they or their suitor end up breaking strictly enforced social conventions.

Today we live in a world replete with diversity. It is a world, as well, where young women have many more choices and possibilities than they did in Austen's day. They can meet an array of people (even on the Internet) whose cultures, assumptions, and backgrounds they may know little about. There is difficulty now, as there was in Austen's world, in judging whether people really are who they say they are or appear to be. A young woman takes the risk that a relationship can sidetrack her own development and desires in the service of a man who in reality only values her as an object and does not share anything very emotionally deep with her. In Austin's world as well as today's, meeting a new man, one whom a girl desires but does not really know, is both thrilling and terrifying. Many of us lead lives that are not local and circumscribed enough for us to know readily or to check whether the people we meet are who they say they are or whether they are good people.

We are not suggesting that Stephenie Meyer is a writer of Austen's level. We are simply suggesting that the sorts of themes Stephenie Meyer and Alex work with do not have to be seen as trivial or childish.

In any case, the excerpt above is a style of writing that is very much one in which Alex has become adept, albeit with her own slant. On her Flickr site, Alex mentions that she read the *Twilight* series in the Fall of 2007. "Lincoln Heights" is obviously inspired by and draws from *Twilight*. These similarities eventually led to a discussion on the SWH site about whether "Lincoln Heights" is original or a derivative from *Twilight*. We will look at this discussion below. Perhaps because of its similarities to *Twilight*, Alex's story began to attract a lot of attention primarily from other teenage girls, the prime audience for the *Twilight* books.

Like *Twilight*, "Lincoln Heights" features an "ordinary" teenage girl, a teenage male vampire, and their romance. The plot twists are similar. The girl and the vampire boy are united, separated, and then reunited. Another teen male briefly attracts the girl's attention. A group of powerful vampire elders attempt to end the romance. These are all plot points that are typical of this genre of teenage vampire romance. While in *Twilight*, the author describes the vampires' beauty in words, Alex is able to use photos to portray increasingly physically attractive Sim characters, characters adored by many of her readers.

Alex posts regular updates on her blog and on the SWH thread to keep her fans informed about progress on upcoming chapters. She posts "teasers" prior to uploading complete chapters. Here is one example:

Lincoln Heights 2.6 Teaser

A frustrated sigh escaped my lips and I charged at the bush, deciding to test just how strong I had become. Leaves and twigs littered the ground at my feet as I tore at the bushes. I gave grunts of effort as I pulled as hard as my limbs would allow, my muscles flexing. Various cuts were bleeding on my face and arms as I lashed at the remainder of the shrubs. But just when I thought I had reached a clearing; my sweatshirt caught on a high branch. At first I turned to untangle myself, but gave up not too long after. There was no getting it back now so I wiggled out of the shirt and leaped through the last of the bushes and into the clearing. (alexvamp15, 2008b)

Alex is, as we have said, making a story from pictures and words. She has to shape both pictures and words to fulfill her desires for her story's design and meaning. From her first story to "Lincoln Heights," Alex has improved in both writing and in the redesign of her photos from the game. A noticeable improvement in her photos occurred, for example, after her mother got her Adobe Photoshop: "I can FINALLY edit my pictures! My cover 4 LH 1.7 looks SICK! I am sooo excited to releaseit!!!! Check bac 4 it! Should b out either this weekend or next!" (MonkeyGalPly, 2008)

Improvement in her writing is also evident. Alex originally had poor spelling, a problem that never goes away completely, but has improved markedly, perhaps because she started to give her stories to editors—other fans who volunteer to proofread—prior to posting them. In fact, proofreading becomes something of a privilege for her readers.

More importantly than spelling, Alex's writing style improves. Compare, for example, the excerpt below from her very first story, to the beginning of "Lincoln Heights" printed following it: "Jordan and his girlfriend, Courtney had been making out for a few minuets. When Jordan pulled away" (MonkeyGalPly, 2007a). The beginning of "Lincoln Heights" (MonkeyGalPly, 2007b) is much more sophisticated. As the story's description,

Alex uses a quote: "'You cannot call it love, for your age the heyday in the blood is tame'—William Shakespeare." The actual story begins as follows (remember that each of these sentences has a picture with it):

> Devin Collins rolled his dark eyes as his younger sister, Julia, whined about leaving all her little friends in New York.
>
> "I still don't get why 'we' have to move here when its Ariel's parents' who died!"
>
> Devin glared at his sister. She was ten: five years younger than himself, but she knew exactly how to piss him off. (MonkeyGalPly, 2007b)

Note here how "whined" and "little friends" are, though in the narrator's voice, actually infused by Devin's voice, as it is he that would think of Julia as whining and her friends as "little." This is a form of indirect quotation that is common in novels. The direct quotation "I still don't get why 'we' have to move here when its Ariel's parents who died!" is not explicitly attributed to the sister, but is clearly understood to be her speaking and, indeed, a representation of her whining. The emphasis on "we" gives the word a stress that sounds like whining.

The introduction of Ariel, a character we do not yet know, starts the story in medias res and, again, places the narration inside the private knowledge and minds of Devin and his sister. This creates suspense and forces the reader to wonder who Ariel is. This, too, is a technique common in novels. Again, "piss him off," though not a direct quote, is from the point of view of Devin. Of course, "rolling his dark eyes" is a phrasing meant to capture teenage girl readers at the outset (and it does—girls on the thread say Devin is "theirs" and comment on his attractiveness repeatedly). Alex, indeed, knows her readers very well.

Character development and characters' physical appearance tend to be important as "hooks" that keeps fans interested in *Sims* stories. Creating attractive Sims was an important skill that Alex continued to develop. Figure 7.1 illustrates her growing ability to create beautiful characters with Photoshop. For Alex's fans, who are mostly other teenage girls, the central male-female couples are often the object of intense interest. Alex played on this interest by asking her readers, for example, to vote on the cutest couple in her stories. Fans often swoon over male characters and get emotionally involved in the stories.

The fan postings got more intense as Alex continued adding chapters to "Lincoln Heights." She began to spend most of her time on SWH but

Figure 7.1 A cover page that Alex created for a chapter from "Lincoln Heights"

Note. Image by Alexandra Salvesen/ "MonkeyGalPly." Originally posted at The Sims2.com Story Exchange (http://thesims2 .ea.com/exchange/story_detail.php?asset_id=251470).

kept getting comments in her Sims2.com guest book (the following quotes are all taken from her Sim Page[3]). Many fans say her stories are "amazing" and some suggest she should publish "Lincoln Heights" as a book. One fan says that "Lincoln Heights" is "THE best story I have seen on this site." Another says "Lincoln Heights" "ROCKS!!" and tells Alex she herself is starting a story soon based on "Lincoln Heights." Many of the forum posts in the "Lincoln Heights" thread are similar to the posts in Alex's Sims2.com guestbook: electronic fan mail. Her fans even call themselves "fangirls" and "story stalkers." Here is a longer quote from a typical piece of fan mail in Alex's Sims2.com guestbook:

ALEX, YOU ROCK!!!!!!!
    Dear Alex,
    I LOVE your story Lincoln Heights . . . Your story has so many twists. I never know what to expect. I check almost every day to see if you put out a chapter. I've read all your storys. I love your writing!!! Seriously, you should think of

publishing it. You'd get milllions. I know it. I wish had Lincoln Heights in book version so that I could read it all day, wherever I go. Lincoln Height is a story I could always relate to! The style, the emotions, and the excitment. It's the best story ever writen!!! To tell you the truth, your like a celebrety to me. Your the best writer ever! Keep up your totaly awsome writing!!!!!

Your Bigest Fan Ever

What is clear is that Alex has a large number of very loyal fans. They see "Lincoln Heights" as a story that speaks to their age, feelings, and experiences in a deeper way than does *Twilight*, in part because it is written by another teen girl. To her fans Alex is a star just like Stephenie Meyer, though she is doing what she is doing off-market, not for money.

Twilight and "Lincoln Heights"

Alex herself eventually invites discussion of whether "Lincoln Heights" is too similar to *Twilight*. This is an important topic; in participatory culture of the sort typical in today's popular culture, fans constantly imitate, play off, and remix published writing, music, and other media. Issues of derivativeness, if not outright plagiarism, arise. On the SWH thread about "Lincoln Heights," here is what Alex has to say about the influence of Stephenie Meyer's work:

Ok soooo i'm gonna b honest and u have to give me ur honest oppinion. U guys know i got the idea about vampires from twilight and i built on after that right? I know i did make it have the same things as twilight (red truck, little town, etc.) but did i really make it THAT much like twilight? The way i look at it i didn't. i just get upset when people say its not original. Hey, its not like Stephenie Meyer made up vampires! (Sorry, Steph, but its true [u know i <3 u!]) but seriously! Arrrrrrrrrrrrrrrrr[.] (alexvamp15, 2008a)

Most replies to Alex's question support her view that "Lincoln Heights" is not just a *Twilight* remake. Alex's question elicited a lot of responses, perhaps because it is an issue (since her story is a derivative) that might pertain to quite a few *Sims* stories (all of the following quotes are from the SWH thread). One fan tells Alex, "You made it your own," and, "Personally I like yours better. More realistic and our age." Another fan says that "yea, you may have gotten SOME ideas from the book" but goes on to say that Alex's story has "tons of twists and turns in there that make it totally original and your own." Another fan says that "Lincoln Heights" is "5 times better than Twilight" and yet another says that it "tops twilight by a mile." Another fan praises Alex as "a true writer" and says she gets "that emotion past the readers, and make them think so deeply."

Some readers, but very few, did not agree with the general sentiment that "Lincoln Heights" was better than *Twilight*. For example, one reader says that "twilight is better in the slightest." But fans rush in to support "Lincoln Heights" after such posts. For example, a post responding to the reader just quoted says that events in "Lincoln Heights" are "constructed a lot better than in twilight" and that "Lincoln Heights" is "a lot more believeable" as a vampire story than is *Twilight*.

Sounding remarkably mature as an author, Alex herself acknowledges that at the start her story was too heavily based on *Twilight*. When someone says, "Lincoln Heights is good an all . . . but, the beginning of Season 1 is a bit too much like Twilight to be able to publish it," Alex writes, "Yeah, it really was. I was new to the genre and I wasn't sure where to go with it so i just stole some stuff from twilight until i could stand on my own two feet. After the first couple chapters i had my own take on it and I finally made it my own. I won't be getting Lincoln Heights as it is published. I would never get it published as it is =]" (alexvamp15, 2009).

In any case, Alex is in no danger of running out of fans or adulation: "I really can't believe that I know someone my age that can write a FREAKIN' AWESOME story and is only 14! Every time I talk to you it's like I'm talking to someone even more popular than Stephenie Meyer or . . . Oprah!"

Another reader hopes all this fame does not go to Alex's head: "Sure, Alex is not quite as popular as Oprah, no offense Alex, but she is really popular on the exchange. I mean . . . can't you believe how many gbs [guest book signatures] entries she gets, with people telling her she's awsome! I think that's really cool! Just don't let all the attention go to your head. Lol."

Twilight Fandom

The Internet and online communities allow amateurs like Alex to have experiences that only a few people used to be able to experience. This is part of what participatory culture is all about. Though Stephenie Meyer is a best-selling author, there is much less difference between her and Alex than we used to. Both Stephenie and Alex have their adoring fans.[4]

Stephenie has a Web site (http://www.stepheniemeyer.com/), as do many famous authors, and so does Alex. Some features of Stephenie's Web site are similar to how *Sims* authors like Alex promote their stories. Stephenie's site has Web pages with brief teasers from each book—such teasers are used by many *Sims* writers, including Alex—and a link to the first chapter of each. Similar to the list of songs that Alex posts with each "Lincoln Heights" chapter, Stephenie has posted a playlist of popular songs that she associates with each of her books.

Stephenie's site includes a list of fan sites in many different languages. On the fan sites, the discussion often seems similar to the fan mail that Alex received. Many *Twilight* fans have begun fan fiction versions of *Twilight*, just as Alex did in the *Sims* universe. In turn, other teens have begun fan fiction versions of Alex's stories.

There is really not a huge difference, other than money, between Stephenie and fan-fiction writers like Alex, or between what Alex's fans and Stephenie's fans write on sites devoted to their stories. In fact, in a special *Time* magazine article, Stephenie's writing is compared to fan fiction, an interesting switch when the author who inspires fan fiction is seen as, in a sense, writing it: "Meyer floods the page like a severed artery. She never uses a sentence when she can use a whole paragraph. Her books are big (500-plus pages) but not dense—they have a pillowy quality distinctly reminiscent of Internet fan fiction. (Which she'll readily grant: 'I don't think I'm a writer; I think I'm a storyteller,' Meyer says. 'The words aren't always perfect.')" (Grossman, 2008).

In turn, in one interview Stephenie refers to fan fiction as inspiration for writing she did after the initial *Twilight* volume. The interviewer asks about the character Edward in *Twilight*. Edward's problem is that he loves his (nonvampire) girlfriend, but she inspires bloodlust in him for her, a desire he must control. The interviewer asks Stephenie if she has ever considered writing a story from Edward's point of view. Stephenie responds,

> Actually, yes. I got started thanks in large part to fan fiction. I didn't even know what that was until someone told me that people were writing fiction about Twilight and posting it on the Web. Naturally I was curious, so I started reading to see what was out there. It ranged from really good to really silly, but one thing all the stories had in common was that they weren't getting Edward right ... No

one seemed to have any idea of how difficult it was for him to live the way he does. It upset me a bit—I felt like Edward wasn't getting enough credit. And then I started thinking about what the first chapter of Twilight would have sounded like if Edward had been the one to tell it. When I finally gave in and started writing, I didn't mean to do any more than the first chapter, but (as so often happens with me) once I started, I wanted to keep going. (BookStories, 2006)

Initially, before her fan base became too large, Stephenie even responded to fans' e-mail questions. One of the biggest *Twilight* fan sites, *Twilight Lexicon* (http://www.twilightlexicon.com/) has a section with Stephenie's replies to numerous questions, many of them aimed at getting more information about characters and events in her novels. It is an intriguing example of a collaboration between fans and author. The *Lexicon* was made by a fan with the screen name Alphie, a fan to whom Stephenie has become close.

Alphie started to write fan fiction inspired by *Twilight*. Her first piece was a story told from Edward's perspective, and Alphie knew that Stephenie was working as well on a version of *Twilight* told from Edward's point of view. Four chapters into her story about Edward, Alphie got a review from Stephenie Meyer herself. Alphie says she was "shocked that she paid me such a nice compliment"(Alphie, n.d.). Alphie asked Stephenie if she could start a *Twilight Lexicon*, to help fans write in *Twilight*'s language. Stephenie said it was an excellent idea and offered to answer any unanswered questions. As Alphie says, "Basically from there on in, if I asked it, she answered it and then some" (Alphie, n.d.).

Interestingly, Stephenie eventually asked Alphie to help her as an editor with the volume *Eclipse* in the *Twilight* series. This relationship is an intriguing example of how fan culture can lead to new relationships among fans and professionals. It also shows the similarities today between some professional writers and fan-fiction writers.

Stephenie has written at length about how and why she came to employ Alphie as an editor (all quotes below are from Meyer, 2006). She wrote about this, in part, because many other fans wanted to work with her and envied Alphie. Stephenie first addresses her fans' concern that the professional editorial process would reshape her stories "in a bad way." Stephenie herself shares this worry:

The professional editors I work with are extremely invasive. They ask for enormous, far-reaching, plot-changing rewrites. Some I give in on, because I can see the merit there. Others I fight . . . Yeah, editing is scary. It's hard. It makes me cry and throw things. I think I'm getting better at it, and hopefully I'm getting a good sense of which changes I should listen to and which I should reject. I never want to get to the point where I think I'm smarter than my editors and I shut them out completely. You can tell when authors do that, and it's not pretty. So editors are both a good thing and a bad thing. (Meyer, 2006)

Stephenie then explains why even though she has four professional editors assigned to different parts of the editing process, she still needs an editing assistant like Alphie. She says that what she needed was "a friend" to help with editing, rather than "distant and busy" professionals. But her fans, many of whom would love to help her, wondered why she picked Alphie.

Stephenie was originally drawn to Alphie's fan-fiction story that rewrote *Twilight* from Edward's perspective, since Stephenie herself was working on the same project. Though Alphie's version was "entirely different" from hers, Stephenie was "impressed with her skill and style." She was also thrilled with the *Lexicon* idea. After Stephenie reviewed Alphie's story and offered any assistance she could with the *Lexicon*, she eventually found e-mail too slow. So Stephenie and Alphie talked on the phone for hours and found out that they "had an amazing amount in common" and "were in sync from the very beginning." Stephenie goes on to say, "I've had a few people morph from 'fans' to friends' and this happened very naturally with Alphie" (Meyer, 2006).

Stephenie says that one of the frustrating things about her job is how long she has to wait for feedback from her professional editors in New York. She was excited about her Edward version and was dying to share it and get feedback. Alphie understood all the difficulties of writing from Edward's perspective because she was doing it herself. Stephenie sent Alphie what she had written of Edward's story, and Alphie sent her "tons of insightful comments—sometimes a compliment, sometimes a question over a confusing point, sometimes a correction when I'd contradicted myself." Alphie would read Stephenie's chapters within hours and get back to her with helpful comments "while the chapter was still fresh" in Stephenie's mind.

Stephenie began to "crave" Alphie's help on the rewrite of her original draft of her book *Eclipse*. The "looming rewrite of *Eclipse*" was keeping Stephenie "up at night" and "the idea of having someone to discuss all the craziness was too appealing to resist."

Alphie does not tell Stephenie what to write. She does not try to get Stephenie to change her plots or characters, as her professional editors sometimes do:

Alphie reads the story. She tells me when something confused her, or someone's eye color was different in another chapter, or when she thinks I expressed something particularly well. She often puts big "HA HA HA HA"s at the end of a line, and I love that :) She also listens while I talk through scenes and plot lines and offers her opinion when I ask for it. I bounce things off of her to get a first reaction. I also whine to her a lot and she doesn't complain. (Our contract does mention that my whining is part of the deal, so she was warned). (Meyer, 2006)

Stephenie's relationship with Alphie raises the question as to whether Stephenie "plays favorites" with her fans. She admits that, "unfortunately," she does. If she wants to have any kind of contact with her fans, "favoritism is inevitable," since she can only answer "the tiniest percentage" of the communications she receives. But she would rather answer some than none, and she would rather visit a few *Twilight* sites than none. Stephenie goes on to say, however, "I adore my fans. I am the hugest fan of my fans. I can't believe people are so excited about my books. It seems like it must be fake, some kind of dream. I want so much to keep in contact with them and enjoy their company when I can, but I can only do this so often, and I can only interact with just that small percentage" (Meyer, 2006).

Reading how Stephenie writes about her fans and her relationship with Alphie exemplifies the close relationship between Stephenie as a professional writer and fan fiction writers like Alphie. Note that Stephenie says that "the professional editors I work with are extremely invasive." Her remarks about professional editors make clear the tension she feels between professionals and amateurs (e.g., fan writers and fan editors). She does not clearly prefer the professionals and treats the amateur writers and fans as experts in their own right. Indeed, she sees interactions with her own fan community as quite important.

In another interesting blurring of fandom and professional knowledge, *Twilight Lexicon* has a section on the book-publishing process, written by a woman who is both a *Twilight* fan and a professional book editor (see Lakegirl, 2009).

Back to Alex

Stephenie and Alex, as writers and readers, professionals and amateurs, are closer than such types of people have ever been before. Borders, once bright, have blurred. People are entering into communities where they tell and circulate their own stories and assume that sometimes the bard, whether professional or amateur, will often sit quietly while the audience tells the story. Indeed, we are all bards, at least potentially.

Age is not crucial here. Just as old women can design homes for little people on computer monitors, even young teens can invite others to hear a story that will both entertain them and give some sense to what they feel and experience. Once again, we see how players' engagement with *The Sims* does not consist merely of playing the game, but extends to using the game as a tool for the creation of meaning.

Alex is a young teen who is learning to write out of school. She is using a multimedia format that requires her to design both pictures and words and make the two support each other. She is getting a unique education, learning to act like a professional with an audience that—just like the professional Stephenie Meyer—she must understand, listen to, and collaborate with. Compare this to the sense of audience and collaboration most young people experience in school. Alex and her audience are using storytelling in a new form to manage the perils and emotions of being a teenager. We know many readers will think different things about Alex and her vampire romances. We, personally, are fans.

Chapter 8

From *The Sims* to *Second Life*

A Young Woman Transforms Her Real Life

Diversity

In the 1960s the United States government decided on something truly new for its schools (Semel, 2002). The country wanted to educate all children, regardless of race or social class. Prior to the 1960s, there was much less concern for the equitable education of poor and minority children. Many poor children did not go to, or at least did not finish, high school and certainly did not go to college. Starting in the 1960s, our society began to believe that all children should graduate from high school and have the opportunity to receive a college education. Many more disadvantaged young people began to take the Scholastic Aptitude Test (SAT; the college entrance test) and, as a result, scores initially went down (Lemann, 2000). The scores eventually rebounded, however. This was a credit to U. S. schools and teachers who took the education of poor children seriously.

Thanks to the continued de facto segregation in our neighborhoods and schools, due to patterns such as "white flight" from urban areas and the clustering of wealthy and upper-middle class in neighborhoods separate from the poor and working class, we still have serious equity gaps in how well richer and poorer children do in school, in terms of high school graduation and in college attendance and graduation (Payne, 2008; Tough, 2008). From the 1960s to the present, scholars in education have devoted immense attention to issues of cultural diversity (Banks & McGee Banks,

2006). In practice, however, this often has meant a focus on race, especially the differences between whites and African Americans, since African Americans are often taken as the iconic U. S. minority group. Attention has been given to class (socioeconomic differences), though much less, and the effects of class have often been confounded with race when scholars treated African Americans as the iconic poor group as well, although there are more poor whites in the United States than poor African Americans.

Of course, in today's twenty-first-century United States, there are many other minority groups than African Americans, and not all people of color are poor by any means. Dozens of different languages are spoken in our schools and streets. By midcentury, whites in the United States will be a minority group, and the country as a whole will be composed only of minorities, with no clear majority population (Bernstein & Edwards, 2008).

The nature of diversity is changing in other ways, as well. Today, thanks to all the different local, national, and global groups and communities that digital technology has made possible, people can take on more "cultures" or identities than they ever have been able to in history. It has been said that Western society has been focused on "self-fashioning" (designing one's own self) since the Renaissance (Greenblatt, 1983), but the trend is certainly at a high point today. People can make and remake themselves in the real world and a great variety of virtual worlds. A white, unemployed, divorced male can be a widely respected guild leader, playing as a woman, in *World of Warcraft*, with lots of status, power, and even *World of Warcraft* gold.

In Chapter 7, we met Alex, an adulated teen writer of *Sims* stories. Alex is biracial (white and Asian). In many parts of her life, this "racial" (or, better, ethnic) identity may be very important to her. But when she is writing her vampire stories and interacting with her fans, she is focused on her identities as a modern digital teenage girl and as a popular *Sims* graphic artist and writer. Educators have stressed racial, class, and ethnic identities but have given much less attention to the myriad of other identities people can create, transform, and recreate today.

Today people are more diverse in more ways than ever before. A great many people today take very different trajectories through life in and out of school—especially out of school—as they engage with different social, cultural, and interest-driven groups and passions.

We said of Jade, the teenage designer in *The Sims* and *Second Life* we met in Chapter 4, that she was typically untypical. Today people who have the opportunity (and not everyone does) and take this opportunity can pursue very different paths to learning, skill development, and identity formation. People today often end up being tech savvy, smart, artistic, and creative in different ways, each of which, however, represents important twenty-first-century skills for our high-tech, complex, globalized world.

As we think about educational reform for the twenty-first century, and wider reforms in lifelong learning, we have to think about people as being able to choose and pursue quite different trajectories to valued ends. The time for cookie-cutter education is long gone. We want to stress this aspect of diversity in this chapter: not racial or class diversity, but diversity in trajectories and choices about identities in all different senses (though, of course, racial and class identities intersect with one's other identities). Here we will introduce a woman who is decidedly typically untypical: Jesse. Jesse is now in graduate school. But in a real sense her graduate education started long ago in spaces that bear no resemblance whatsoever to school.

Jesse and Virtual Worlds

Jesse today is in her early thirties (we have changed her name for privacy and withheld some identifying information). As a child, she had a supportive extended family, was a good student in elementary and high school, learned a lot about computing from her older brother, was a talented high school athlete, earned a college degree, and held and still holds a demanding job at a state university. She is biracial, white and Hispanic. All this may sound pretty standard. However, it was Jesse's experience in virtual worlds that, in her words, helped her recognize her talents and discover what she wanted to do with her life.[1] It was in virtual worlds, too, where Jesse saw diversity in new terms.

Virtual worlds are two-dimensional (2D) or three-dimensional (3D) online environments where people come together for a wide range of purposes such as just "hanging out," playing games, having virtual sex, operating a business, designing things, or engaging in political activism. Some virtual worlds take the form of massively multiplayer online role-playing games (MMORPGs), with *World of Warcraft* as the most well known and popular example. In such games, hundreds or even thousands of people can be playing and interacting with each other (usually in small groups) at the same time. *World of Warcraft* is a game where, together, people accomplish quests, or tasks that players complete in order to gain new skills, acquire valued items, and so forth. Not all virtual worlds are games; some are just environments where people can interact with each other and engage with a variety of different activities. (There is a rapidly growing literature, both scholarly and popular, on virtual worlds. Some examples include Au, 2008; Bartle, 2004; Castronova, 2005; Dibbell, 2006; Taylor, 2006.)

There is a myriad of virtual worlds in existence today for people of all ages. Worlds designed for younger participants—such as *Club Penguin*, *Whyville*, and *Toontown*—have more restrictions on the kinds of interactions possible among residents to reduce the likelihood of inappropriate behavior and language. In virtual worlds for adults, while the companies that own them do restrict certain behaviors, participants must still sometimes negotiate a good deal with each other about norms for language and behavior.

Most virtual worlds allow for considerable customization of content. For example, in *Gaia Online*, residents can choose and decorate their own home, import content, and modify content ranging from objects to ground coverings. Most virtual worlds have their own currency, currency that can often be purchased for real dollars or earned within the world. In turn, the virtual currency can often be converted back into real dollars. As we saw when we discussed Jade, people in *Second Life* can earn Linden dollars—*Second Life*'s currency—and Linden dollars can be exchanged for U. S. dollars. Thus, people can make real money in virtual worlds like *Second Life* (see Castronova, 2005, and Dibbell, 2006, for interesting discussions of virtual economies).

At least for adults, *Second Life* (SL) is one of the best known virtual worlds. *SL* was launched by Linden Lab in 2003 with the goal of providing participants with a community-driven and user-created experience. While *SL* has had its ebbs and flows in participation, recently it hit a record number of more than eighty-eight thousand residents online simultaneously, with about seven hundred fifty thousand unique users logging in every month (Linden, 2009).

The vast majority of *SL*'s content is created by its residents, using built-in editing tools as well as importing content created and modified with other software, such as Adobe Photoshop. Residents can purchase virtual land, a feature that has spawned a booming real-estate market, including developers who buy empty lots, landscape them, and even build homes on them, selling the improved lots for a hefty profit to *SL* participants who do not want to learn to build for themselves. Resident-owned businesses sell everything from clothing to fully furnished houses. Popular items include "skins," consisting of virtual hair, makeup, and even body shapes that surpass the customization possible with the built-in tools to edit one's avatar (i.e., in games or virtual worlds, a person's representation of him- or herself).

The decision to allow residents almost complete freedom in designing the world's content, as well as in their choice of activities, has had a huge impact on *SL*'s development. Linden had to create a system for identifying and separating areas with graphic sexual content and residents on the adult world must be eighteen or older. There are separate islands for teens, where they are protected from unwanted contact and content (at least content unwanted by their parents and other adults).

Moving around the *SL* world can be difficult for newcomers. Residents must master the techniques of flying and teleporting, since there is no organized grid of roadways or paths. Despite increasingly improved tutorial orientation experiences, newcomers typically have a hard time getting acclimated to the *SL* movement system, interface, and tools. In addition, the unstructured nature of the world leaves many people wondering what to do next. While there are many interesting places to visit and resident-organized events, it is up to the individual to decide where to go, what to do, and whom to interact with. Further, the in-world tools for building content can be intimidating for people unfamiliar with digital-editing tools, scripting language, or solid geometry.

Despite these challenges, over the course of *SL*'s existence, residents have created a fascinating mishmash of social groups, identities, norms, and practices set within a landscape that ranges from exquisite tropical gardens and stunning architectural reproductions to sleazy dance halls, glitzy shopping malls, and jarring billboard advertisements. Residents can design their avatars in any way they like and there is a massive diversity of avatars, human and nonhuman looking. They can act out any gender or personality they wish.

Not surprisingly, *SL* has attracted considerable attention from the popular press, though often these stories focus on the seamy and controversial aspects of *SL*, such as the virtual sex industry, virtual crime, and residents who become so enamored of *SL* that they abandon their real-life relationships and jobs. A more balanced and complete picture of the richness and complexity of having "a second life" is, however, documented in a growing number of publications by *SL* residents and researchers (some examples include Au, 2008; Boellstorff, 2008; Meadows, 2008; Rymaszewski, 2007).

Educators have become interested in the potential of virtual worlds such as *SL* as environments for learning. Many universities now have campuses in *SL*. Linden Lab actively recruits educators to *SL* (see Linden Lab, 2009), and, over the last few years in particular, educational activity in *SL* has increased tremendously.

Initially, many educators seemed to view *SL* primarily as a way to recreate traditional classroom instruction, with the advantage of bringing together students from diverse locations. It was (and still is) common to see virtual classrooms in *SL* equipped with desks, lecterns, and even PowerPoint presentations. Now, however, educators are starting to take advantage of the 3D environment to create more immersive, interactive learning experiences, including simulations of real-life habitats, architectural replications, language immersion experiences, and many other creative applications (see Jo Kay, n.d.).

Interestingly enough, as educators move into *SL* to create educational activities, the educative potential of *SL* as an environment in its own right, without any formal courses, tends to be overlooked. *SL* has its own social and cultural landscape, and its own history, residents, traditions, and tools. We argue in this chapter that life in *SL*, and other virtual worlds, can be an education in itself. It certainly was for Jesse. We need to give more attention to how virtual environments create learning opportunities that may be far more powerful than even the most engaging 3D-simulation classroom designed by an educator.

Another point we will make in this chapter is that life in virtual worlds cannot and should not be viewed as separate and distinct from life in the "real world." As Jesse's story will show, the real and virtual are now melding in unpredictable and potentially beneficial ways. Jesse is a real graduate student in a high-tech area—a very skilled graduate student, indeed—but her virtual experiences, by and large, were excellent preparation for her future learning as a graduate student. In a real sense, though she did not know it, her graduate education started in the virtual worlds into which Jesse stepped long before she went to graduate school.

Jesse's Story

We first met Jesse when we were both teaching a summer-school course at the university where Jesse works. Jesse was introduced to us by her mother, who worked at the university and heard we were teaching a course on literacy and learning in games and virtual worlds. Jesse's mother thought that Jesse's experience as a gamer, and particularly her involvement in *Second Life*, might be of interest to us and to the students in the course. We kept in touch with Jesse after the course was over and interviewed her several times for this chapter.

Jesse was and still is employed at the university in a very demanding staff position. She handles daily issues related to building maintenance and space allocation for one of the university's colleges, ranging from ensuring broken windows are fixed to overseeing faculty and staff office relocations due to construction projects. Another aspect of her job is managing the logistics of college events, large and small. She also is the frontline technology support person for her college, handling routine information-technology (IT) issues before referring them to the campus IT support office. Jesse attributes much of her "base technology knowledge," as well as her confidence with computer technology, to her gaming experience.

Jesse has an undergraduate degree in human resource development, and she worked in retail for a year before she obtained her current position, which she has held about seven years. When we asked her why she had not pursued a degree in computer science, she said that she followed her brother's example. He became an accountant because he wanted games and computing to be "fun" and not an extension of his job.

Jesse gave a guest lecture in our course and sat in on a number of sessions. She told us a year later that her experience in the course was "life changing" because she realized for the first time that her knowledge and experience with games and virtual worlds might be valuable to others. She was invited to do several guest lectures in other courses at the university after the summer session.

Six months after she spoke in our course, Jesse started a master's degree program in information and learning technologies at a local university. This program is offered entirely online, which does not faze Jesse given her previous experience in online environments. She is taking courses such as "Designing and Teaching in eLearning Environments" and "Developing eLearning Instruction." Her gaming experience has already been useful in her coursework. She is developing an in-world course, "Introduction to *Second Life*," as a major project. She also noted that the skills she developed as a disc jockey (DJ) in *The Sims Online* (which will be discussed shortly) were useful for streaming music in an online presentation she created for a course last spring.

Jesse's Background as a Gamer

Jesse attributes her interest and initial exposure to games to her older brother (who is eight years older than she is). Her brother worked for their aunt and uncle, who owned a computer business. He put games on their home computer for her to play, and she recalls playing games starting at around age seven or eight. Her parents, in contrast, encouraged "outdoor activities." She says that her mother "still doesn't get it" when it comes to her and her brother's interest in gaming and feels left out when they talk about games.

Jesse started playing *The Sims* when she was an undergraduate in college. She played *SimCity* as a hand-me-down from her brother and thought *The Sims*, created by the same company, would be fun. She played *The Sims* through the release of many expansions, ending with *The Sims 2: Pets* in 2006. She enjoyed building the most in *The Sims*, but she also found it "cool" to run Sims families through their simulated lives. She did not use many cheats (codes to make life in the game easier), since she liked to experiment and "see how the game would play out."

Jesse also did not participate in any fan communities and says she is not sure that she knew that any existed. She did talk about the game with her brother and with some cousins who also played it. However, unlike the women we discussed previously, she did not create custom content (content not available in the game as it comes), perhaps because she did not visit online fan communities and thus was not aware of the game's modding potential.

Playing *The Sims Online* (*TSO*), a massively multiplayer online game variation on *The Sims*, was Jesse's first experience with a virtual world. *TSO* was launched in 2002 and received rather mediocre reviews, never attracting the sizeable audience that was predicted, based on the popularity of the single player *Sims* games. Much of *TSO* was similar to the single player *Sims* games, except that players controlled only one avatar (not a family) and could interact with other players in thirteen different virtual cities. *TSO* continued for quite a few years with a relatively small membership, until it was closed and relaunched as the free *EA Land* in 2007.[2] *EA Land* was speculated to be a competitor to *SL*, but it lasted only a few months before being shut down.

Jesse abandoned *TSO* long before, in about 2004. One reason that *TSO* was never as popular as *The Sims* or *SL* is that players' ability to create content was very restricted. Certainly Jesse later found *SL* much more appealing because of the opportunity it gave her to be a designer and creator.

Jesse's brother bought her *TSO* for Christmas in December 2002. He had quite a bit of prior experience with online multiplayer gaming and cautioned her about the potential for sexual harassment online. He convinced her to create a male Sim avatar and to pretend to be a male who was playing the game, which she did. Jesse's male gamer persona had a personality similar to her own, but she made up a different job and family attributes (such as having a sister rather than a brother). Later she found out that many other *TSO* players had done the same thing for privacy (though many of her friends were upset by her deception when she revealed her true gender later on, in *SL*).

Interacting with strangers online was uncomfortable at first, but Jesse remembered getting to know some people in a "skill house" where they collectively were developing their Sims' skills. Even though she initially found it a bit strange to interact with people online, the game required her to engage in social interaction to "skill up" her avatar and make money. Eventually Jesse estimated that she got to know sixty to seventy people as players in *TSO*, some who are still her friends today (she estimates that she has met about four of these friends face to face). Much of her early game play was exploratory and fun. She and her friends had parties, chatted with text chat, and "got married a lot." She eventually became an online DJ and played music for two hours at a time. Often she'd log into *TSO* after work and play from about 6 p.m. to 10 or 11 p.m.

Jesse's friends in *TSO* introduced her to *SL*. She played *TSO* for about a year and a half before starting *Second Life*, played both for about a month, and then switched over entirely to *SL* because most of her friends were now there. She recalls her amazement at the difference between the 2D environment of TSO and the much more compelling 3D world of *SL*. The 3D environment added a tremendous appeal to her interactions with other people in *SL* in the virtual world.

Jesse has played a couple of MMORPGs while she has been involved with *SL*. She started playing *EverQuest II* and thought it was great. The combination of *EverQuest* and *SL* was too demanding for her computer and so she stopped playing *EverQuest* when some other friends quit. Since last summer's class, where we discussed *World of Warcraft*, she has been playing that game a good deal. It is clear that Jesse's trajectory has been much influenced by friends she has met online. In fact, her online social network is now massive. We are used to the influence that peer groups have on young people. But today, peer groups are spread across the world and are made up of people of many different ages, often playing different sorts of people in virtual worlds than they are in real life.

Formal Education

Jesse received a good education in school. She also received a good education, as we will see, in virtual worlds. The two sometimes complemented each other, but the virtual worlds' education eventually predominated.

Jesse grew up and went to school in the same community through high school. She was a good student (on the honor roll) in elementary and middle school and an average student in high school. She spent much of her after-school time from elementary school through high school playing sports. In high school, she played varsity volleyball and softball, earning two state championships in volleyball. She received an athletic scholarship from the University of Minnesota, but decided not to pursue college sports due to an injury she experienced as a high school sophomore. Though she no longer plays competitive sports, she plays intramural softball two nights a week. She also boxes for exercise, a sport she picked up in college. Jesse said she makes a point of keeping physically active, something she neglected at times while she was heavily into *SL*.

Jesse described her initial experience as an undergraduate in college as a huge cultural shock. At the time she attended, the university did not have, as she put it, a very (racially and ethnically) diverse student population. She had a racist roommate in her freshman year, and in general as a biracial woman, she felt that she was not fully accepted by either white or Mexican American students on campus.

Jesse eventually left her dorm room during freshman year and (unbeknownst to her mother) moved in with cousins and commuted to campus. Though, in her junior year, the campus began to make significant efforts to improve the college climate for minorities, on the whole Jesse had a difficult time as an undergraduate. She graduated from college with a degree in human resource development. She chose human resource development because she liked helping people and was interested in finding ways for businesses to recruit and retain people of color.

Jesse's story is a good example of the complexity of diversity in today's world. In college, she did not feel accepted as a biracial woman. She would, however, find massive acceptance in virtual worlds, though as a person with many different identities beyond her biracial identity: virtual identities, such as designer, DJ, spouse in virtual marriages, and virtual parent, she has shaped and fashioned for herself as she interacted in these worlds. Some of the people Jesse has met in virtual worlds know who she is in the real world and accept her not just as a biracial person, but as a person with a whole set of other identities. As we will see shortly, one woman who Jesse met online has come to live with her in the real world. Jesse is also, among her many other identities, a lesbian.

Here is what Jesse said about diversity in one phone interview: "Well the way I see life is that everyone you encounter can create a learning experience for you and the more diverse group I can encounter, the more I'm going to learn. I'm surrounded by diversity just in my own family in real life, but it doesn't compare really to the amount of different points of view I've encountered here [in *Second Life*]."

Second Life

Every time we interviewed Jesse, *SL* was present in her real life. During one interview, a friend from *SL* was visiting her. At the start of another interview, she told us she had just been working on a project for a graduate course and had found a video of a news program that highlighted a *SL* business person whom she knew well. And at a third interview, the woman who was soon planning to move into her real-life household joined us in *SL* for the interview.

Jesse started spending time in *SL* in 2004, three years after she graduated from college and two years after she began her current job. At that time, she was playing *TSO* and friends from *TSO* introduced her to *SL*. The greatly enhanced ability to build things in *SL* was what engaged her from the start. Two friends taught her how to manipulate basic shapes. She then began taking building classes offered by the Linden staff. She still recalls the first class, "Building 101," where she learned to drill holes in objects and make a table with glass in the center.

She took a number of other classes and did a lot of experimentation on her own. For example, she recalled learning how to create a spiral staircase out of just one prim (a prim is, in essence, one building block in *SL*, so this was a very complicated task) by looking at the interior construction of such a staircase that had been created by another player. For the first three to four months, she estimated that she spent nearly every day after work in *SL*, learning to create different things.

When we asked Jesse what she found so appealing about building in *SL*, she said that she never was a very "creative" person, and yet, suddenly, in *SL* she could be creative. In contrast to manipulating a 3D model on the computer, she found it compelling to be able to get inside of what she built (with her avatar), walk through it, view it from different angles, and make changes with immediate results. And, as she said, there was no need to wait months for a builder to remodel a home, as in real life. The gratification was immediate. She had liked to play with Legos as a child and she felt that *SL* building was an extension of that, but it allowed her to go far beyond. Designing in *SL* is a demanding technical and aesthetic task, and Jesse has now become a skilled designer in that world.

One of the intriguing things that Jesse mentioned in our summer course was how she learned geometry through building in *SL*. She had been good in algebra at school but was (in her words) terrible at geometry. She attributes her mastery of geometry in *SL* to several factors. One was the potential described previously, that is, to interact with what she created from different perspectives. Another was the ability to see the "numbers" simultaneously with the actual 3D model; in other words, to see how the numeric measurements corresponded to actual angles, shapes, and so forth.

The *SL* building tool displays the parameters of even simple objects (see Figure 8.1). These parameters include the object's position, size, and rotation, all in three dimensions (that is, on x, y, and z axes). Related to this factor, and perhaps most importantly, according to Jesse, was that she learned to pay attention to and manipulate these numbers while creating objects in *SL* rather than focusing solely on the shape itself. By watching the numbers change as she manipulated objects, she gained what learning scientists (Clark, 2008; Gee, 2004, 2007) call a "situated" and "embodied" understanding of, for example, how the angles of each part of a roof need to match up. Jesse claimed that she knows many people who were not good with geometry in school who have become adept builders in *SL*.

This sort of learning is, in many respects, exemplary whether done in or out of school. First, using geometry to build things as part of a valued community can make geometry meaningful, valuable, applicable, and

Figure 8.1. Screenshot of Hayes's avatar and the building tool in *SL*

important. If students in school had such opportunities, perhaps they would pay much more attention to their geometry lessons (or lessons on other subjects where a similar approach could be taken) and maybe even need less of them. Second, situated and embodied understandings—understandings where people have seen abstract knowledge applied to specific contexts and in a way that recruits their bodies, even if virtual bodies—are much deeper and longer lasting for human beings than abstract, decontextualized learning and learning based on telling and texts alone (Barsalou, 1999; Clark, 2008; Gee, 2004, 2007; Glenberg, 1997).

Jesse's "Family"

At one point in our conversations, Jesse mentioned that she came to *SL* with her "family." When we asked her what she meant, she told us that it was common in *TSO* for people to form families, with a couple serving as parents and offering to "adopt" other players. Jesse explained that her *TSO* parents were a couple who owned a house in *TSO* that was a popular hangout for her and lots of other players. Jesse began calling them "Mom" and "Dad," since they were nurturing people and seemed to fill that role for many of their *TSO* friends. They offered to "adopt" her, and she became their first child (to be followed by many others).

Eventually, Jesse began to participate in what she called a *TSO* "Mafia." This is a phenomenon that has been written about a good bit, even in mainstream publications like the *Boston Globe* (Bray, 2004). As this article and Jesse described it, Mafia groups engaged in quite a bit of "griefing," or troublemaking, in *TSO* by doing things like "tagging" (making someone your enemy, which lowers their popularity in the game and makes it more difficult to make new friends) and taking over someone's lot and destroying their possessions. When *TSO* players migrated to *SL*, the Mafia migrated with them, though Jesse's group evolved into a more benign form.

Groups like the Mafia in *TSO* (and in *SL*) are controversial, of course. People in virtual worlds often differ on what the norms for behavior ought to be. Is it all play, with people doing everything they want and the game or world allows? Or should they follow rules of behavior that are more similar to the real world? There is a famous episode in the *World of Warcraft* where a guild was holding a funeral for a member who died in real life (Davies,

2006). An opposing group attacked them and disrupted the funeral. Many found this deed deeply reprehensible. Others said that *World of Warcraft* is a game that allows and encourages warfare—that is the purpose and rules of the game—so the behavior was just fine. The point is that people in virtual worlds often have to negotiate over what is acceptable talk and behavior and they do not always agree.

For the first six months or so that Jesse was in *SL*, she resisted joining in any Mafia activities there. Her "sister" from *TSO* had created a huge Mafia family in *SL*, mostly comprised of former *TSO* participants. After repeated urging from her *TSO* and *SL* friends, Jesse finally joined her sister's *SL* Mafia family. These Mafia families assumed the names of famous Mafia families, like Costello and Genovese.

To gain rank in the Mafia, members had to spend a lot of time in *SL* carrying out Mafia business. Jesse ultimately became an "underboss" with responsibility for training "soldiers," spending six to seven hours after work online. Her Mafia family grew to more than one hundred members. Jesse estimated that there were at least five other *SL* Mafia families with memberships ranging from ten to one hundred fifty members. As she explained it, she was part of a group of four families that banded together to fend off opposing families.

Eventually her sister got tired of spending so much time on her Mafia duties and, as that family disbanded, Jesse formed her own Mafia family with someone else. Their family grew to about forty members, whom Jesse still calls her "kids." Her relationship with her kids does sound like parenting. She felt that she was an important source of emotional support and advice (about real-world issues) for them. Many of them were teenagers (this was before Linden strictly enforced the age requirement, so a number of younger teens managed to gain access to *SL*). She said that she was able to convince several of them to stay in school in real life.

As Jesse described it, many of these kids found *SL* more compelling than the real world because they were more popular or able to assume identities that were "cooler" and more attractive than their real-life identities. In fact, Jesse said that she felt somewhat the same way. She felt like she could be more "herself" in *SL* than in real life, more outgoing and social, particularly as a single woman who was not inclined to hang out in bars.

Jesse estimated that she spent three years in *SL* Mafia activities, with one of those years as head of her own Mafia family. According to Jesse,

Mafia activities died out in *SL* several years ago. Jesse's Mafia family by and large interacted with other Mafia families and engaged in social interactions with each other, rather than griefing people who did not play or want to play the Mafia "game."

Jesse felt she learned a lot from her Mafia experience about leadership, dealing with interpersonal conflicts, and problem solving. Clearly, what stood out for Jesse about the experience was not any thrill from engaging in illicit activities (in fact, she says that Mafia activity in *SL* was pretty tame compared to *TSO*). What was central for her was the experience of being part of a family, of being a role model for younger players, and of managing the daily interactions of forty or so people in her own virtual family. Even now, she maintains relationships with some of her kids in *SL*. In fact, and, perhaps, oddly, Jesse's Mafia activities taught her social and emotional intelligence.

Unlike many other *SL* players, Jesse did not get a job or own a business in *SL*, though she says that many of her children did (one owned a wedding business). She wanted to "keep the game a game," though most of us would consider the time and effort she spent in the Mafia as much like work as any job.

Jesse's real-life family is changing in lots of ways due to her *SL* experience. Her *SL* virtual "wife," Misha, is moving in with her in the real world. In the past, Jesse had a *SL* friend move in, with bad consequences, including wiping out her savings. She's been more cautious this time.

She and Misha have been "married" in *SL* for close to four years. Marriages are common in *SL*, but typically dissolve quickly. Jesse had been married three times in *SL* before she met Misha. Their relationship has been through some trials, particularly when Misha discovered that Jesse was not a man in real life, as she had pretended to be. The move is a huge commitment on both sides. Misha is moving from another state with her three kids from a prior relationship. Her family is not comfortable with the idea of a lesbian relationship and she tends to describe Jesse as a friend.

Jesse seems excited by the impending change in her household. She's moving her boxing equipment out of the basement so they can set up beds for the two boys. About two months ago she cut back on her time in *SL* drastically because she needed the time for her graduate work. She gave up her own property and only logs in occasionally to visit her "kids." She said, too, that Misha helped her realize long before that she needed to set boundaries on her relationships with her *SL* children, since they could usurp all of her free time. Misha's two boys love to play *World of Warcraft*, and Jesse expects that she will be in *World of Warcraft* with them more often than *SL* once they move in. She does not think she will give up her avatar or her *SL* relationships altogether, but *SL* will play a much different role in her life.

Real or Virtual?

When we met Jesse in *SL* for the first time, we were not quite prepared for how striking her appearance would be. Jesse's avatar was a large, well-muscled, very attractive man, with dark skin and black hair. He was dressed in all black, with silver jewelry. Jesse said he was "five years in the making."

Jesse was obviously very comfortable and adept in the *SL* environment. We teleported to a public "sandbox" where anyone could build (on private property, owners can restrict building privileges to prevent unwanted construction). Since there was considerable "lag" (the amount of time between when you make a command, like moving your avatar across a room, and how long it takes for the avatar to actually move), Jesse created a wooden platform for us and moved it up in the air, where there was much less lag. When asked how she made the platform "fly," she explained that she simply set the "z" axis in a way that repositioned the platform upward (remember, she now knows geometry).

Jesse had arranged for her *SL* and soon-to-be-real-life partner to join us. Her partner was still in her home state, putting up signs for her garage sale, packing her belongings, and otherwise preparing for her upcoming move to be with Jesse. While we waited for her, Jesse began showing us pictures of some *SL* houses she had built in the past. We were simply astonished by the size and beauty of her houses. The first one she showed us was only the third house she had ever built. She built it during the time she was highly active in the Mafia and it was designed to hold forty people (the maximum that could "fit" on a parcel at that time). She jokingly commented that "we didn't do much but hold meetings all of the time." The second house, also from her Mafia days, had a huge wraparound porch.

The third house that Jesse showed us was called Da Mansion and the house was, indeed, a mansion, with an enormous swimming pool. Jesse estimated that it took her about six hours to build the house. It was based on blueprints she and Misha had found online. These were blueprints for a real house that Jesse then translated into *SL* measurements and constructed, at the same time designing her own additions that made the house considerably larger than the original blueprint. Jesse is such a good builder in *SL* that she could make a lot of money building for others. But, as we said, she wants to avoid having her gaming become "work."

Jesse's partner finally joined us. She was very funny and pleasant and engaged in a lot of banter with Jesse. She obviously thinks very highly of Jesse and described her as someone who continually underestimated her own abilities. Jesse pulled an entire building out of her inventory and "rez-zed" it (this is the equivalent of taking a tiny box out of your pocket, putting it on the ground, and opening it up, allowing an enormous building to appear). We went into the building, while Jesse tried to remember when she built it and for what purpose.

We then got into a conversation about another house that Jesse built. When Jesse built this house, she and Misha decided to have triplets in *SL*. There are user-created tools in *SL* that allow players to simulate an entire pregnancy and birth. As Jesse and Misha explained it, you buy tummy attachments in different sizes that simulate the gradual growth of pregnancy. You also buy "baby whispers" that are messages given off by the pregnant tummy at different stages, regarding the mother's cravings and feelings, as well as the baby's. The player decides how long the pregnancy should last, and then delivers the baby in an *SL* doctor's office.

Jesse pulled out a baby Sim to show us. It was adorable, sitting on the floor and playing with a musical toy. Our conversation then shifted to the *SL* residents that Jesse and Misha have "adopted" as their *SL* children. Many of them, but not all, were involved in the Mafia. They estimated that they currently have seventeen children (some of whom have their own children, and Jesse referred several times to her grandchildren), though the number has ranged from about five to forty. Jesse said that "raising a family is the true work I've done here," referring back to the teens she described in an earlier interview, but also to the general work of managing the relationships among young adults who were her kids.

Jesse's partner had to get back to packing and her real-life children, so she left, and we asked Jesse to show us one of her favorite places. She teleported us to Apollo, a gorgeous island that Jesse uses as a sort of retreat.

We arrived at a spot with tai chi scripts (scripts are programs that *SL* participants write, in this case allowing avatars to do tai chi moves), lots of landscaping, bird songs, and calming music. She said that she had been coming to the island for about four years, and that it had endless places to explore. We began to talk about the way that *SL* has allowed people to express themselves artistically, in ways often not possible in the real world. Jesse mentioned as an example an island that one of her (virtual world) granddaughters had shown her. The island told a story in poetry and visual images, a story that allowed multiple interpretations.

Jesse went on to tell us that there was something that no one in her real life knew about her—that she sings. She began singing as a DJ in *TSO*, after friends heard her sing on the phone and liked it. They "forced" her to sing the next time she disc jockeyed and she continued in *SL*, singing on every show. She said that not seeing her audience, and knowing they could not see her, helped her get over her anxiety. She has, however, never sung solo in real life, only singing in her choir.

It was getting late, and as a final question we asked Jesse if she would miss *SL* (as you recall, she is spending much less time in *SL*), since she had had such rich experiences there. She said, "No, because the best things that came from Second Life are taking me places in my real life . . . I'm getting my masters degree finally and I actually know what I want to do with my life . . . and of course Mel and the kids are moving here." We said that we hoped her story could educate other people. She replied that she hoped so too, and went on to observe, "A lot of great things can come from gaming, you just have to keep your eyes open to the opportunities as they arise and not continue to stay focused on the game once they're upon you." We have talked a lot in this book about gaming beyond gaming, that is, the many activities gamers do beyond just playing their games. Jesse's remark points to the future of gaming: one should not stay focused just on the game when the game offers one other opportunities for learning and development.

Jesse: Typically Untypical

Jesse is biracial. She is a lesbian. She was the head of a Mafia family. She had lots of kids in *SL*, some of them adults. She is a DJ. She is a superb builder and designer in *SL* where she is held in high regard for her technical skills and her artistry. She has a massive social network in both the virtual world and the real world. Her diversity is not captured by saying only that she is biracial and lesbian, the sorts of identities to which we tend to pay the most attention when thinking about diversity.

Jesse, similar to the other girls and women we have discussed in this book, picked up high-tech skills together with artistic skills and skills associated with social and emotional intelligence. These skills, so often separated in our talk of school reform today, especially in our obsessions with STEM science, technology, engineering, and mathematics (STEM), are not separate in these women's lives and development. These skills most certainly should not be separate in school and in society as we face the complex problems of our globalized twenty-first-century world.

There is a story we hear frequently in talk about equity and schooling today. A minority student, perhaps even more marginalized by sexual orientation, faces oppression and turns off from and eventually drops out of school. Jesse's school success waned in high school and she faced such oppression in college. But she was engaged in another educational track altogether. This track was outside of school. It did not look like "academic" learning. But the skills Jesse learned—technical skills, math skills, social skills, organizational skills, emotional skills, artistic skills, design skills, networking skills, and communicational skills—are all readily transferable to school, the workplace, and society. They are, especially in their integration and combination, a true set of twenty-first-century skills.

Today, Jesse is a graduate student in a technical field—information and learning technologies—but a technical field that requires social and emotional intelligence, since it requires thinking about how humans learn and perform best with technologies and with each other. Jesse came to this program, and to her current job, where she is now helping to organize and run learning and technology conferences, well prepared indeed. But a good deal of this preparation did not come from school. It came from activities in virtual worlds that look very little like school. For Jesse, these activities were life changing. They were a real graduate education before she started graduate school. She woke up one day and realized that her activities in virtual worlds had made her highly skilled and valued in real life.

Many readers will find at least parts of Jesse's story strange. She seems so diverse and "out there." She lives in virtual worlds and brings them into her real life. She refuses to be any one thing or to let other people dictate what counts as important types of diversity and identity for her. She does not look and act like one's privileged child, who, perhaps, is a good student headed to an elite college. What she has done seems, perhaps, so much stranger.

But, believe us, we who have interacted with Jesse, you do not want Jesse as competition in the twenty-first century. She is too skilled in too many ways to compete with. It took her time to find herself; after all, she had so many worlds she had to move through. Her pedigree is not what we used to take as typical for success in the world. But that is because today, more and more, to succeed, you need to be typically untypical, that is, you need to be one of the great number of people today who are pursuing their own unique trajectories of learning and life.

There is nothing standard about Jesse. But then, "standard" does not get you much in today's highly competitive, high-tech world with its many different niches, identities, and cultures. It will get even less in the future as millions and millions of well trained Chinese and Indians, and many others, compete for jobs, resources, and status across the world.

Jesse's story tells us that we really do not know fully what learning looks like in the modern world, and we have lots of misconceptions about the matter. Today learning takes many forms and we need to start paying attention to them. People are learning valuable skills in many cases that, at first glimpse, look trivial, odd, like simply messing around, or merely social. Things look this way to us because school learning has left such a strong stamp on our minds. Much valuable, even high-tech learning does not look like school today. People across history have always learned a great deal informally. But today, in games, virtual worlds, and other digital settings, people are learning quite formal and transferrable skills in organized settings that are educational without looking like education as we usually conceive it.

Virtual worlds can be dangerous places, as Jesse discovered when a virtual world friend stole from her in the real world. Indeed, today parents and educators are obsessed with Internet safety to the point of banning many sites (e.g., social-networking sites), technologies (e.g., mobile phones), and software (e.g., text messengers). We are so sure we know what learning looks like that we cannot imagine social-networking sites or text messaging could have much to do with it. But we do not know really what learning looks like in today's world.

The real world can be a dangerous place as well, but we realize that people have to learn to handle it, be savvy and smart, and still take some risks. Humans learn from experience, and virtual worlds have added a great deal to the experiences we can have, experiences that are not really isolated from the real world. Jesse has had a lot of experiences that most of us have not had. She refashioned herself. Now she knows what she wants to do in the real world. She is, indeed, untypical, like the other girls and women in this book. Untypical will be the new typical in developed countries in the twenty-first century.

Chapter 9

What Does It All Mean?

What Women and *The Sims* Have to Teach Us about What Education and Learning Will Look Like in the Twenty-First Century

Schools and School Reform

Today, in the United States and many other developed countries, we face the twenty-first century with out-of-date schools (Wallis & Stepoe, 2006). Too many of our schools focus on information and facts in an age when these are all cheaply available on the Internet. They fail to focus on problem solving.

Too many of our schools focus on standardized skills in an age where people with only standardized skills will be competing against lower-cost competition in China and India (Friedman, 2005). They fail to focus on innovation.

Too many of our schools focus on what students know now in an age where skills, information, and technologies quickly go out of date. They fail to focus on preparation for future learning (Bransford & Schwartz, 1999).

Too many of our schools focus on preparing students for jobs in an age where most jobs are service jobs and do not pay well or bring people much status (Reich, 1992). They fail to focus on preparing students to develop skills for off-market activities and communities that will bring many more people status and satisfaction (Leadbeater & Miller, 2004).

Too many of our schools focus on individual achievement in an age where almost all real problems, and most high-tech workplaces, demand skills in team work and collaboration. They fail to focus on students' abilities to engage in knowledge-building networks with smart tools and other people (Bransford, Brown, & Cocking, 2000; Gee, 2004, 2008).

Too many of our schools underutilize technology and are, indeed, frightened by it as authorities ban Internet sites, mobile devices, and games in an age where almost all deep learning recruits technology. They fail to focus on making students as digitally literate as they are (or should be) print literate (Gee, 2004).

Too many of our schools focus on knowledge as a noun—the things people know and believe—not as a verb—the things people can do with knowledge. They fail to focus on knowledge building and design thinking in an age where, outside of school, the trend is for people, young and old, to produce and not just consume (Bereiter, 2002; Jenkins et al., 2009).

Too many of our schools treat students as consumers, and often passive ones at that, in an age when young people produce, design, modify, and make choices in their popular culture. Schools fail to focus on producing proactive learners with ties to learning communities that mentor them and allow them to mentor others.

There are many different approaches to school reform today, emanating from government, business, nonprofits, foundations, and university schools of education. Many of these reforms still focus on tests and test passing, rather than problem solving and innovation. Too many reflect a mania for STEM (science, technology, engineering, and mathematics)—technical and technicist education—cut off from social and emotional intelligence, ethics, and the arts, all of which are the foundations for making intelligent use of and decisions with technology. Too many reformers have failed to see the importance of collaboration and sustaining learning communities both for learning in school and for out-of-school, lifelong learning.

We in the United States are obsessed with the fact that students in countries like China beat our students in science and mathematics on international tests (Lemke et al., 2004). But in a complex globalized world with environmental (e.g., global warming), economic (e.g., worldwide poverty), and civilizational (e.g., terrorism) problems that all interact with each other, just knowing algebra is not enough. We need students who know algebra but can also pool that knowledge with other people's knowledge, smart tools, and technologies and apply it with social and emotional intelligence and ethical thinking.

Learning Today

Schools have competition today of a sort they have never had before. All sorts of institutions and communities today are using digital technologies (computers, the Internet, mobile and handheld devices, and game platforms) to produce learning every day. Many young people today are learning all the time, whether they are in school or playing a game at home. In fact, many of them are learning more complex things, more important twenty-first-century skills, and more technical skills at home or in their communities than they are at school (Gee, 2004, 2007; Shaffer, 2007). Many older people are learning every day as well, often off-market in communities on the Internet.

And yet learning, at least highly valued learning, does not just happen by itself. Well-off parents resource and mentor their children to use technology, including games, for production, knowledge building, and innovation. In the terms we used in Chapter 2, they cultivate their children (Lareau, 2003). They have been doing this for years with traditional print literacy, with reading, and with writing. Today they are doing it with digital literacy, that is, with digital tools for making meaning.

We need to worry about the children who do not get such resources and mentoring. These are children who may have access, at home or in a library, to digital tools like game machines and digital music players, but who do not get rich mentoring to challenge themselves, persist past failure, and continually develop new skills (Neuman & Celano, 2006). These children certainly do not get these resources and mentoring at school for the most part. Today, if they get them at all, they get them from community programs, libraries, after-school programs, or Internet communities.

While we most certainly need to reform schools, schools should not be expected to do the job all by themselves. Well-off children today are using digital technologies to learn in and out of school. They network with diverse people, young and old, at home, in their communities, and on the Internet. Poorer children cannot be left with an education only in school and education can no longer be equated just with school. School reform must cease to be "school reform" and become "learning reform," that is, the reform of learning as an everyday, multi-institutional, global, lifelong concern. School must be seen as part of a bigger picture and integrated with out-of-school learning. Otherwise, poor children will never be able to compete, and our schools will continue to be test-preparation academies for standardized skills that are in copious supply across the world.

Learning Reform: What Women Gamers Have Taught Us about Learning and Literacy

In this book we have looked at girls and women playing games and going beyond gaming to produce, design, and write, especially in *The Sims*. We have done this because some of our previous work concentrated on more action-oriented video games and what they have to teach us about learning and literacy. This earlier work (e.g., Gee, 2003, 2004, 2007) paid no special attention to women gamers, ignored *The Sims*, and gave little attention to the activities gamers do beyond game play and to the communities in which they often do those activities. We wished to redress that limitation and draw on our other, more recent work to generate new insights into learning (Gee & Hayes, 2009; Hayes, 2005, 2007, 2008; Hayes & Gee, 2009, in press; Hayes & King, 2009).

Our purpose here is still to learn lessons about learning, this time from women "gaming beyond gaming." The lessons we can learn cannot give us the whole picture of what school reform becoming learning reform needs to look like in the twenty-first century. But these lessons can give us some important parts of the picture. Furthermore, these lessons can give us parts of the picture that have been less discussed and focused on in work on games and learning when women who go beyond gaming have not be the center of attention.

Simulations

We have learned about the power of simulations like *The Sims* and *Second Life* (a world where some of our *Sims* players have ended up). *The Sims* is for entertainment, and so it is not a realistic simulation of life, or certainly of poverty. Yet in the third chapter we saw Yamx in her *Nickel and Dimed* challenge create (and negotiate) a set of rules that tweaked *The Sims* software to allow players to simulate the life of a poor, single parent, not in the sense of being completely realistic, but in the sense of creating the feeling of what a poor, single parent faces in life.

Yamx created new rules of play by thinking about a piece of software and about an important social problem and how the software could be made to represent the problem in a way that led to empathy and understanding. She mentored a community into existence that wrote about their game play and reflected both on poverty and the nature of simulations like *The Sims*. She created an entertaining game that was, at the same time, a socially conscious game. Yamx allowed players to learn new skills, both technical skills and creative ones in their writing. She showed that *The Sims* was not a scientific or even "good" simulation but that this fact could be made into a virtue.

Yamx integrated the cognitive understanding of poverty, the social understanding of poverty, and an emotional understanding based on emotional intelligence that stressed feelings and perspectives. She melded game play and gaming beyond game play, since group discussions and the albums the players designed became an integral part of their activity.

Simulations are at the cutting edge of science today. Complex systems, such as the environment, the global economy, the clash of cultures and civilizations, and the spread of viruses, among a great many other such systems, systems that play such a big role in our global world, are not open to traditional experimentation. Too many variables interact in too many complex ways. Scientists use simulations to model such complex systems. They watch what happens when they run the simulation and compare it to reality. Then they redesign the simulation to get closer to reality. Finally, as the simulation gets more accurate, they seek a general understanding or theory of what the simulation tells them about reality. Such work almost always requires different specialists to combine their knowledge. This is not that far removed from what Yamx was doing, albeit for fun.

One important difference between game simulations like *The Sims* and scientific ones is that players "play" (control) a character (or sometimes more than one) in game simulations (as the player controls her Sims in *The Sims*). The player is, in a sense, both outside and inside the simulation. This can give rise to an important educational effect if used properly. Players can be encouraged to combine two perspectives, a top-down, big-picture consideration of the simulation as whole and the insider perspective from the point of view of the character (or characters) they control.

This juxtaposition of outside and inside perspectives can be powerful. It was what allowed Yamx and her players to consider both the nature of *The Sims* as a poverty simulator and the nature of poverty in the real world, on the one hand, and the specific challenge of being poor in a given context within the game. Precisely because *The Sims* is not a great simulator of poverty, Yamx and her players had to tweak it and think about how to make it better. They had to think, as well, about what it meant to simulate something like poverty at all.

Educators in and out of schools can use entertainment simulations like *The Sims, Second Life,* and many others in much the same way Yamx did to create a learning community reflecting on important problems and ways of understanding today's complex realities. With today's digital technologies, we can create simulations of almost anything. Learners can start with game simulations, like *The Sims, SimCity, Civilization, Zoo Tycoon, Spore;* these simulations encourage learners to think about rules of play and how they do or not reflect reality.

Learners could move on from the game simulations, which are motivating because they are focused on play and entertainment, to more scientific but still game-like simulations. For example, learners could start with the commercial game *SimCity* and then move on to more professional and authentic, but still game-like, urban planning simulations, such as the one constructed by David Williamson Shaffer (2005, 2007) at the University of Wisconsin–Madison. In Shaffer's simulation, called *Urban Science* (Bagley & Shaffer, 2009), students replan part of their own home town inside a *SimCity*-like game world but use real urban planning tools and in the end have to construct a real planning report for real planners. Finally, students could move on to real professional urban-planning tools and activities wholly in the real world. Such tools, of course, involve statistical models and various tools to graph and represent reality.

This progression could happen in any area. For example, *Quest Atlantis* at the University of Indiana is a game-based science simulation (Barab et al., 2007). It puts learners in a game-like world where they have to solve science problems collaboratively. These scientific problems are ones that are also caught up with social, political, economic, and environmental issues.

For example, one problem requires students to figure out the source and type of river pollution in a park where there are competing interests (e.g., park visitors, fishermen, loggers, and the park's need to make money). What the students propose to do about the pollution must take into consideration a variety of interests and not just scientific facts.

The point is not that students do not need to master scientific and mathematical skills. They most certainly should. But in the twenty-first century, they need to know more than this. They need to know how to integrate these skills in collaborative problem solving where knowledge is shared and distributed across people and smart tools and technologies. They need to know how to use these skills in problem-solving contexts where social, ethical, cultural, economic, and political issues also apply, as they always do in the real world. Finally, they need to know how to use these skills in complex problem solving contexts where they recruit social and emotional intelligence so that they can change people and not just things. Simulations, employed in much the way Yamx used *The Sims*, are a promising avenue here.

A future educational agenda, one already initiated by the serious games industry, is to build simulations of various degrees of professional fidelity that allow students to engage with complexity, to meld both technical skills and social and emotional intelligence, and to recruit both scientific and artistic thinking for discovery. This is just what Yamx did. Surely professional educators can do so too and in a wide variety of areas.

Passion and Production

In Chapters 4 and 5 we saw that today people, young and old, want to produce and not just consume. In Chapter 4, the teenager designer Jade developed technical and design skills through *The Sims* and *Second Life*, not through school. Jade's new skills were part of an identity transformation. She came to see herself as a confident, tech-savvy producer, designer, creator, and even an entrepreneur when she sold her creations in *Second Life*. Tabby Lou, shut-in and in poor health in her sixties, found a true second life as an adulated designer for *The Sims* and a mentor within a lively learning community of the sort we called a passionate affinity group. Neither Jade nor Tabby Lou wanted just to consume or just watch others produce.

They wanted to produce themselves and they wanted to have audiences themselves. This is a major trend today in popular culture, a trend that affects both young and old.

School is largely about consumption. Students consume the knowledge and beliefs others have produced. They rarely produce knowledge themselves. Knowledge works very differently inside and outside school. Outside school, whether in scientific pursuits or in popular culture, knowledge is about what you can do with it. It is about what problems you can solve or how you can change things for the better. Inside school, knowledge is about storing facts in one's head and writing them down on tests. The learning scientist Carl Bereiter (2002) has referred to these two orientations to knowledge as a *belief mode* (school), where what is important is what others have claimed or believed, and a *design mode*, where what is important is whether knowledge and beliefs are adequate to tasks we need to carry out and how knowledge can be put into use. A belief mode is important, but it is inert when not combined with a design mode.

When students are learning science, mathematics, social science, or anything else in school, they rarely get to mod the curriculum, design things for the curriculum, and produce knowledge themselves. These are all things that many young people do get to do, and expect to be able to do, in their popular culture. Surely, such proactive production would aid in real problem solving and innovation, two major aspects of preparing students for the twenty-first century.

Science and other knowledge-building enterprises primarily engage in what has been called "model-based thinking" (Hestenes, 1999; Lehrer & Schauble, 2006). Models are simplified representations of real objects or systems where the representation is similar to some ways to the object or system it represents. Think, for example, of a model plane used either in play or in a simulated wind tunnel for scientific tests. Think, too, of various sorts of diagrams, graphs, maps, equations, spreadsheets, statistical models, and blue prints. These are more abstract models than a toy plane, but they are models nonetheless. Simulations (which these days are often virtual worlds) are just large and intricate models that seek to represent relationships in a system.

Scientists almost always use models to understand reality. They build and manipulate their models because they usually cannot so readily and safely manipulate the real world. The real world is often too complex to understand all at once without a simplified model. Models constitute a design aspect of science. They have to be produced and assessed (and often aesthetic criteria like simplicity and elegance count in science when evaluating models). Models and model-based thinking, when used in education, introduce elements of design, production, and modification for students if they are allowed to build, tweak, and evaluate and redesign models the way scientists do.

Keep in mind that simulations, like *The Sims* or more academic ones, are just big models. Just as a toy plane is a model of a real plane, *The Sims* is a toy version of reality. All models are, in a sense, toys, since the real world is too complex to take on all at once, and we use models to simplify parts of reality enough to get a handle on understanding it.

There is, however, one key problem with any education that becomes centered on problem solving, using and building knowledge, mastery, and innovation. Such deep learning does not come cheap. It requires mastery of skills, and this mastery requires thousands of hours of practice (Ericsson, Charness, Feltovich, & Hoffman, 2006; Gladwell, 2008). Expertise, of the sort the girls and women in this book have, requires a great deal of practice. You can memorize Newton's laws of motion in a few minutes. It takes thousands of hours of practice to become expert at applying them. While we do not need our students to be experts in the sense that professional physicists are experts, nonetheless real comfort with building and applying knowledge in any domain requires a great of practice.

To get hours of practice requires persistence. Learners must persist past failure and frustration. They must continually challenge themselves to move up the hierarchy of skills in a given domain, and this means yet more failure and frustration even for advanced learners (Bereiter & Scardamalia, 1993). Of course, gamers know all this very well. No one plays video games without failing at times and even failing later in the game when new and harder challenges are thrown at the player. But, interestingly, real gamers do not want games to be easy. They revel in failure and challenge and overcoming them. Even when they do not revel, when they are frustrated and on the verge of throwing a video game controller, they persist. So do scientists. So do the girls and women we have watched playing and designing.

As we have said in earlier chapters, persistence requires passion, a drive for learning and mastery. Passion plus persistence we have defined as grit (Duckworth et al., 2003). Grit is, in our view, the key twenty-first-century ability, since no deep, life- and identity-changing learning happens without it. The girls and women we have studied in this book all have grit in their gaming and designing (even if they do or did not have it in school).

Every child, indeed, every human, has the right to know what it is to have a passion that leads to persistence and mastery. Children need, as well, to know how to recruit and grow passions in the future across their lifetimes. Grit—passion plus persistence—is key for students who are going to learn problem solving and innovation. Grit is also crucial for adults who may need to find status, power, and a sense of control and worth off-market, in deep interests shared with others and not in things like service work.

By and large, we do not associate school and passion, though we often associate popular culture and passion (as we have seen with the girls and women in this book). Today, when young people pick up passions, including those that meld technical skills with other skills, they do so out of school. One thing that kills passion in school is that there is so little room for choice. People choose their passions; they cannot be forced on them.

Another thing that kills passion in school is that everyone is expected to do and know the same things. This is a cherished belief in the United States and many other countries. We are concerned about what every American needs to know about math, science, or civics. Too often, this cherished belief about what every American should know about a subject leads to most of them knowing very little.

In popular culture, young people pick specialties—some are into *The Sims*, others into digital music, and others are into fan fiction, among other things—and pursue them passionately. They often, as we have seen, join communities where, as they advance in expertise, they teach and mentor others. It is not uncommon for young people to collaborate with others who have different specialties to pull off larger projects like a large game mod (where expertise in graphics, audio, game mechanics, and programming are all needed).

We could imagine schools encouraging every young person to develop a specialty that they can also teach to others. They would also be required to work in teams with young people with other specialties to complete larger tasks. Working this way would require the growth of a good deal of common knowledge as each student learned enough about what the others knew to be able to collaborate successfully. This is, indeed, how so-called cross-functional teams work in high-tech workplaces (teams where each member must have deep expertise, but must be able to understand each other member's expertise well enough to integrate and collaborate with them; Parker, 2002).

Schools will always, we suppose, have to mandate learning that some students would not otherwise choose to learn. We hope, of course, that in such cases, the school makes clear what this learning really means and why it is important. Showing how it can be put to use and combined with things the student cares about is one way to do that. Thus, we think it is important that schools also give all students access to resources for developing passions outside of school. Sparks of interest that individual students develop, which cannot always be fanned for the whole class, should always have a possible outlet for development out of school. In-school and out-of-school resources, learning opportunities, and learning communities (whether at home, in libraries, in community centers, or out-of-school programs) need to be closely integrated. However, there is one wrinkle that has yet more important implications for learning in and out of school. We do not know a lot about how passions form and develop. We know even less how to help people cultivate them. We do not know how passion combines with persistence to form grit and really allow the learner to thrive.

In this book, we have suggested that, often, passions grow from small, seemingly insignificant things (sparks) based on a deep desire to do something. To accomplish the desire the person needs help, often a tool. Very often this help or tool is available from a community with a passion (a passionate affinity group or, at least, an interest-driven community) on the Internet (or, in some cases, in the real world), a community that supplies mentoring. Eventually the person becomes hooked on the community and its passion, seeing the community as a place where a trajectory toward mastery can occur. The person now has a shared passion in a site of belonging and continuous development and learning.

A great deal of learning and knowledge production today is done as part of various types of interest-driven groups and learning communities (Ito et al., 2008). One type of these we have called passionate affinity groups (see also Gee, 2004, 2008). In the future (and it has already begun today), schools will be in intense competition with out-of-school learning communities. Educators will need to know how to build these affinity groups in school as well as how to bridge to and use out-of-school communities.

Passionate Affinity Groups

People are learning and gaining passions in all different types of interest-driven groups on the Internet and in real life. These groups build and maintain powerful tools for design, production, and creativity. They offer mentoring and the chance for status in the group and an audience if one has grit.

Not all interest-driven groups behave the same way. Some are tough on newcomers and engage in sarcasm and flames. Some have a very strong sense of insiders and outsiders, like many cultures in the world do. In Chapters 5 and 6, we looked at two communities devoted to *The Sims* that had a set of features that led to particularly healthy learning and development. We called communities with these features "passionate affinity groups."

These passionate affinity groups were organized around a passion their participants wanted to spread. Socialization on the sites was deeply important but was organized around the shared passion for design and *The Sims*, not done just in and for itself. They were welcoming to newcomers and incorporated a good deal of diversity, young and old, new and expert, and everything in between.

These groups honored tacit knowledge while creating many opportunities for people to learn to explicate their knowledge. Members readily combined their individual knowledge with the knowledge other members had and with the knowledge built into the many smart software tools they used (distributed knowledge), and they often engaged in collaboration. They were open to knowledge from other sites (dispersed knowledge), such as the many software-related sites that could help with their passion for design.

These groups encouraged people to specialize and develop deep expertise in a given area, but they all shared a good deal of common knowledge about *The Sims* and design. They took a very proactive view of learning that did not rule out, but rather encouraged, asking for help. Leadership was porous and flexible in the group, as people led or followed and mentored or got mentored at different times. Everyone was an audience for other people, while expertise earned one a large audience of one's own.

Finally, many of the experts in the group, the people like Tabby Lou who had earned massive audiences, did not see themselves as self-sufficient. Rather, they saw their expertise as a product of the community as a whole and embedded in that community. Their knowledge was part of something bigger, and there was always something new to learn.

There is no doubt that interest-driven groups, even quite harsh ones, are the driving force of learning, knowledge production, and innovation in today's popular culture (Ito, et al., 2008). People are becoming such experts in these groups that they are beginning to compete with professionals. Experts in these interest-driven groups are sometimes called professional amateurs (pro-ams; Anderson, 2006; Leadbeater & Miller, 2004). These groups and the learning they are driving is entirely an out-of-school phenomenon. Yet the way they organize learning is, in many respects, much deeper and longer lasting than much of the learning that goes on in many schools.

The features of passionate affinity spaces seem to lead to both deep learning and a healthy and inviting community. They are the models for what schools must seek to accomplish both inside school and in the links schools make to out-of-school learning. It is one thing to learn geometry as an individual because the school tells you have to (it is part of what "everyone ought to know"). It is another to be part of passionate affinity group using geometry to create three-dimensional (3D) designs in *Second Life*.

We are not saying we should stop teaching geometry, physics, and biology. We are saying that such teaching can be preparation for learning in interest-driven groups (at their best, passionate affinity groups) that apply this knowledge to design and production in a socially motivating context. This would mean that the geometry, physics, or biology would take on a more pressing meaning. The student would need it, and the student's work in the interest-driven group (where there was some choice) would motivate them to learn more and pay more avid attention.

This is not different from project-based learning. It is just that, today, project-based learning can be much deeper than ever before. It can be tied to an interest-driven group—ideally a passionate affinity group—that is sustained over time, contains new and more expert learners, accumulates knowledge, mentors newcomers, and allows them eventually to mentor others. There is no reason all schooling needs to be age graded. It is entirely good that students new to geometry, for example, get to be part of a group that can mentor them and pass on to them a whole set of values and norms about design, production, knowledge, skills, and creativity that involve geometry.

Becoming a Star

We have seen that today's popular culture allows both young and old people to gain wide audiences and become stars in their own domain. In Chapter 5, we met Tabby Lou, an older woman with millions of fans, and, in Chapter 7, we met Alex, a teenager writer with a devoted following. In the past, popular culture was based on a big audience watching the professional star. Today's popular culture creates a great many subaudiences and allows non-professionals to have audiences of their own, though often off-market, that is, not for profit (Ito et al, 2008; Jenkins et al., 2006).

We saw in Chapter 7 that the professional writer Stephenie Meyer, author of the *Twilight* books, feels close to her audience, especially to those who are writers of fan fiction around her books. She feels camaraderie with them that she seems not to feel with her professional editors. Writer and audience bear some deep similarities to each other; they are each, in fact, writers.

Alex, of course, feels a good deal of camaraderie with her audience, too: her fellow teenage girls. She and they are very much on the same wavelength in a way that an adult could never be. In a certain sense, Alex is ministering to her fellow teens by telling then a story that helps them make sense of their feelings and lives. She is telling them a story that they can tell themselves, as some of them start their own fan fiction based on Alex's work. Alex has a real standing and role in the community for which she writes. Whatever we think of the feminist politics of vampire romance stories and the stories Alex tells, at the level of sharing with her audience, there is nothing inauthentic about what she is doing.

The sorts of communities in which Alex shares her writing and her reflections on her writing are not fully passionate affinity groups as we have defined them (not every interest-driven group needs to be). Most of the people in these communities, reading and responding to what Alex is writing, are teenage girls. These communities have a tendency to sometimes lose sight of the writing, which is the ostensible focus of the community (its passion) and to become largely social sites. Some writers will then sometimes complain that the focus is being lost. But a compelling story or writer, like Alex, will bring many back to their passion for stories.

It is obvious that writing and learning to write at school works very differently than it does for Alex and her *Sims* stories. First of all, Alex is designing a multimodal text; that is, she is designing graphics and words to go together and to communicate both separately and together. Second, she has a real audience that she must please, listen to, communicate with, and, at times, serve as a part of their audience when she reads their stories. Third, her writing has meaning and value within her shared culture with the other teenage girls reading her work. These girls are culturally diverse (Alex is herself Anglo Asian), but they share concerns, issues, and emotions as twenty-first-century teenage girls. This is a "culture" or cultural identity, too.

Writing in school is usually done on command for no true audience other than the teacher. Neither the students nor the teacher usually care very passionately about the topic and the writing rarely connects in deep ways with some culture the student shares with other students or with any audience for which the writing is written. This may be the reason that many students write so little and like writing so little in school.

In an age when young people can garner audiences and become stars by writing or producing other things (e.g., music, graphics, video, or games), school faces stiff competition (Black 2008; Lam & Rosario-Ramos, 2009). Too often in school, the only way to be a star is by being a good student, which means performing only for the teacher as an audience.

At the same time, it is clear that school could both create writing communities and link to ones out in the world beyond school. These communities could have some of the features of the sorts of communities for which Alex writes. They would not have to involve just fiction (David Shaffer, mentioned above, has another simulation where students become science journalists), though some should certainly, in this day and age, involve multimodal texts. They would have to involve some degree of choice. A group choosing to write in a particular way about a particular theme would have to care about the theme, and the writing would have to speak to their own cultural interests, issues, feelings, and fears. (Keep in mind that we are using "culture" broadly here, not just as racial or ethnic diversity, but as all the different shared social identities people have, such as being a teenage girl or a anime fan.)

It would surely not be a good thing to grade the writing (or other productions) in such communities. In part, the communities, when well organized, hold members to high standards and are already making good and valid judgments of quality. The writing in these communities could be seen as preparation for future learning in later more school-based tasks and assessments. Teachers could give graded writing assignments that tested how much knowledge had transferred from writing in the community to writing in other settings. In such ways, grades would not warp the integrity of the communities.

We could imagine the day when students in school are working in all sorts of interest-driven groups, some of them passionate affinity groups, others more restricted (e.g., in terms of age), in and out of school. All this work would be seen as important learning in its own right and as preparation for future learning in other, perhaps more academic, settings. School would support learners in their work within these interest-driven groups and would assess the transfer of a great many skills from this group work to other more school-like and workplace-like settings. For example, Alex clearly did not learn to spell in school, but learned much better spelling by having an audience that demanded it. Clearly, she would now do much better on any spelling test than she would have done before. At a deeper level, we could use school to help her transfer her skills with multimodal design (graphics and words) to other settings that involve multimedia, graphic arts, design, and writing, and we could assess this transfer.

Diversity and Learning Trajectories: Becoming New Kinds of People

Jade (Chapter 4), Tabby Lou (Chapter 5), Alex (Chapter 7), and Jesse (Chapter 8) all became different kinds of people—found new identities and senses of themselves, as well as new networks, relationships, and audiences—through digital media used in interest-driven communities. Deep learning always requires and leads to a change of self. We begin to see ourselves in new ways, gain new powers, and contract new sorts of relationships with others. In this sense, learning out of school is often deep in a way it is not in school for many students. At the same time, we know that, for some students, school can be deep in this life-changing way. It ought to be so for all students.

The learning landscape in society has vastly changed. People can learn in a variety of different settings using a variety of different digital tools. All sorts of institutions and organizations, some for profit, some not for profit, some formal, some informal, are supplying learning and learning tools today. The Internet, mobile and handheld devices, game platforms, iPhones, iPods, and a great many other devices are greatly increasing our opportunities for learning and changing the ways in which we learn. But tools are only part of the story. Communities—interest-driven groups, passionate affinity groups among them—are the other part of the story. These groups allow people to put digital tools to use to develop shared passions, shared passions that transform their skills, relationships, and identities.

People today, young or old, can take very different trajectories through the new learning landscape and develop themselves in very different ways. Jesse, whom we met in Chapter 8, took a very special trajectory to her current skill set, career aspirations, and identity as a graduate student. She guided a family of teens and adults in a virtual world. She married there and even had virtual triplets. She learned to use complex 3D design tools to design buildings. All of these experiences were not just "play," with no relevance to the so-called real world. Jesse was not within a "magic circle," (Salen & Zimmerman, 2003, p. 95) cut off from the real world. She now lives in the real world with one of her former virtual spouses. She also has an amazing set of real-world technical and artistic skills, as well as leadership skills, and social and emotional intelligence. There is no need to give her a transfer test to see if she can transfer her skills in gaming and virtual worlds to the real world. She transferred them herself.

Jesse, Jade, Alex, and Tabby Lou all seem untypical. What can we learn from such untypical cases? They are special. But we have argued that for those who want to succeed and survive in a developed country in the modern world, you had better be special; you had better be untypical. Untypical is the new typical.

As people pursue their own special trajectories through the learning landscape (made up of tools and groups), they redefine diversity. Jade is a rural white teen. Tabby Lou is an old, shut-in, white woman. Alex is a biracial teen. Jesse is a biracial adult lesbian. But this is not even close to representing all they are. They have many other cultures or identities (or whatever you want to call them). Jade is a successful designer and entrepreneur. Tabby Lou is a designer and mentor. Alex is an adulated writer and artist. Jesse is a high-tech expert, leader, organizer, and networker. They are all other things as well and will be yet other things in the future.

They are expert learners prepared to continue to grow and develop, no matter what their age.

Schools, which now stand so separate from the rest of the learning landscape, will have to integrate with other means and locations of learning. Schools will have to offer a variety of learning trajectories and integrate with others out of school to offer a variety of powerful identities and their concomitant skills and powers.

Girls and women, gaming and gaming beyond gaming in *The Sims*, look, perhaps at first glance, unimportant. They are just playing around. They are not in school, they are not working, and they are not doing something "real." Those are our traditional criteria for what activities are important. But what is going on here is an important part of our future if we are to surpass the many risks and problems in today's globalized world. It is not, by any means, the whole story, but it is an important part of it. We need to learn to work together to reshape ourselves as ethical, emotionally intelligent, lifelong learners in a complex, high-tech world. The girls and women we have met are good guides here, indeed.

Notes

Chapter 3

1. All quotes are from The *Nickel and Dimed* Challenge Forum (BBS) thread (Yamx, 2008). No message or page numbers are available. Unless otherwise noted, all quotes are from posts by Yamx.

Chapter 4

1. The forum posts cited throughout this book are transcribed as they were written. These posts are often grammatically incorrect and include ASCII symbols and emoticons. Readers unfamiliar with this language may even find some terms incomprehensible, such as "lol" (laugh out loud). This is a form of Internet language that is quite typical of how teens and many young adults communicate on these forums and in other digital contexts. This language evolved, in part, from a need to communicate quickly (thus the condensed language), to express emotions overtly, and to convey one's identity as an "insider" within various Internet communities.

Chapter 5

1. All direct quotations from Tabby Lou and other women in this chapter are from interviews conducted by e-mail in 2008.

Chapter 6

1. A humorous example: a digital interactive curriculum on *Moby Dick* made at the Massachusetts Institute of Technology, available on the Web, could not be used by many schools because they banned any sites with the word "dick" in them.

Chapter 7

1. As note 1 in Chapter 4 indicates, the forum posts cited throughout this book are transcribed as they were written and reflect a common form of language used online. We have used the term "teen-speak," since there is not a commonly used descriptor for this type of language; however, it is not confined to use by teens only.
2. There are, however, critiques of this view of romance and women readers; for various perspectives, see Christian-Smith, 1993; Krentz, 1992; Radway, 1991.
3. The complete link to Alex's Sim Page is http://thesims2.ea.com/mysimpage/pod.php?user_id=2877919.

References

alexvamp15. (2008a, April 1). Re: Lincoln Heights [Message 44]. Retrieved from http://similik.proboards.com/index.cgi?board=stories&action =display&thread=3126&page=3

alexvamp15. (2008b, August 30). Lincoln Heights 2.6 teaser [Message 280]. Retrieved from http://similik.proboards.com/index.cgi?board=stories &action=display&thread=3126&page=19

alexvamp15. (2009, January 18). Re: Lincoln Heights [Message 381]. Retrieved from http://similik.proboards.com/index.cgi?board=stories &action=display&thread=3126&page=25

Alphie. (n.d.). The Lexicon story: According to Alphie. Retrieved from http://www.twilightlexicon.com/?page_id=12

Altick, R. D. (1957). The English common reader: A social history of the mass reading public, 1800–1900. Chicago, IL: University of Chicago Press.

Amazon.com. (n.d). *Twilight* (The Twilight Saga, Book 1) Editorial Review. Retrieved from http://www.amazon.com/Twilight-Saga-Book-1/dp/ 0316015849/ref=sr_1_1?ie=UTF8&s=books&qid=1249223046&sr=1-1

Anderson, C. (2006). *The long tail: Why the future of business is selling less of more.* New York: Hyperion.

Andrews, E. L. (2008, October 23). Greenspan concedes error on regulation. *The New York Times.* Retrieved from http://www.nytimes.com/ 2008/10/24/business/economy/24panel.html

Anonymous. (n.d.). *Nickel and Dimed* challenge. *Cuckoo for Cocoa Puffs Asylum* [Web site]. Retrieved on December 20, 2009, from http://www .freewebs.com/sims2fan08/nickelanddimedchallenge.htm

Au, W. J. (2008). *The making of Second Life: Notes from the new world.* New York: HarperCollins.

Bagley, E. S., & Shaffer, D. W. (2009). When people get in the way: Promoting civic thinking through epistemic game play. *International Journal of Gaming and Computer-Mediated Simulations, 1*(1), 36–52.

Baker, E. L. (2007). 2007 Presidential address: The end(s) of testing. *Educational Researcher, 36*(6), 309–317.

Banks, J. A., & McGee Banks, C. A. (2006). *Multicultural education: Issues and perspectives* (6th ed.). Danvers, MA: Wiley.

Barab, S., Zuiker, S., Warren, S., Hickey, D., Ingram-Goble, A., Kwon, E.-J., Kouper, I., & Herring, S. C. (2007). Situationally embodied curriculum: Relating formalisms and contexts. *Science Education, 91*(5), 750–782.

Barker, L. J., & Aspray, W. (2006). The state of research on girls and IT. In J. M. Cohoon & W. Aspray (Eds.), *Women and information technology: Research on the reasons for under-representation* (pp. 3–54). Cambridge, MA: Massachusetts Institute of Technology Press.

Barker, L. J., Snow, E. S., Garvin-Doxas, K., & Weston, T. (2006). Recruiting middle school girls into information technology: Data on girls' perceptions and experiences from a mixed demographic group. In J. M. Cohoon & W. Aspray (Eds.), *Women and information technology: Research on the reasons for under-representation* (pp. 115–136). Cambridge, MA: Massachusetts Institute of Technology Press.

Barsalou, L. W. (1999). Language comprehension: Archival memory or preparation for situated action. *Discourse Processes, 28*(1), 61–80.

Bartle, R. (2004). *Designing virtual worlds.* Indianapolis, IN: New Riders.

Barton, D., & Tusting, K. (Eds.). (2005). *Beyond communities of practice: Language, power, and social context.* Cambridge: Cambridge University Press.

Bereiter, C. (2002). *Education and mind in the knowledge age.* Mahwah, NJ: Lawrence Erlbaum Associates.

Bereiter, C., & Scardamalia, M. (1993). *Surpassing ourselves: An inquiry into the nature and implications of expertise.* Chicago: Open Court.

Bernstein, B., & Edwards, T. (2008, August 14). An older and more diverse nation by midcentury. *U.S. Census Bureau News.* Retrieved from http://www.census.gov/Press-Release/www/releases/archives/population/012496.html

Black, R. W. (2008). *Adolescents and online fan fiction.* New York: Peter Lang.

Boellstorff, T. (2008). *Coming of age in Second Life: An anthropologist explores the virtually human.* Princeton, NJ: Princeton University Press.

BookStories (2006, August). *Interview with Stephenie Meyer.* Retrieved from http://chbookstore.qwestoffice.net/fa2006-08.html

Bransford, J., Brown, A. L., and Cocking, R. R. (2000). *How people learn: Brain, mind, experience, and school: Expanded edition.* Washington, DC: National Academy Press.

Bransford, J. D., & Schwartz, D. L. (1999). Rethinking transfer: A simple proposal with multiple implications. *Review of Research in Education, 24*(1), 61–100.

Bray, H. (2004, January 14). Justice has its price in Sim world. *Boston. com News.* Retrieved from http://www.boston.com/news/globe/living/articles/2004/01/14/justice_has_its_price_in_sim_world/

Brown, A. L., Collins, A., & Dugid, P. (1989). Situated cognition and the culture of learning. *Educational Researcher, 18*(1), 32–42.

Brown, J. S., & Thomas, D. (2006). You play Warcraft? You're hired! *Wired, 14*(04). Retrieved from http://www.wired.com/wired/archive/14.04/learn.html

Brown, L. R. (2008). *Plan B 3.0: Mobilizing to save civilization (substantially revised).* New York: Norton.

Castronova, E. (2005). *Synthetic worlds: The business and culture of online games.* Chicago: University of Chicago Press.

Centre for Educational Research and Innovation. (2008). *Innovating to learn, learning to innovate.* Paris: Office for Economic Co-Operation and Development Publishing.

Chi, M., Feltovich, P., & Glaser, R. (1981). Categorization and representation of physics problems by experts and novices. *Cognitive Science 5*(2), 121–152.

Christian-Smith, L. (Ed.). (1993). *Texts of desire: Essays of fiction, femininity and schooling.* Oxford: Routledge/Falmer.

Clark, A. (2008). *Supersizing the mind: Embodiment, action, and cognitive extension.* Oxford: Oxford University Press.

Daily Mail Reporter. (2009, March 26). Family-friendly Nintendo Wii becomes fastest-selling console in history. *MailOnline.* Retrieved from http://www.dailymail.co.uk/sciencetech/article-1164941/Family-friendly-Nintendo-Wii-fastest-selling-console-history.html#ixzz0NEQ2ANMr

Darnell, M. S. (n.d.). Games based movies suck. *Gametour.* Retrieved from http://www.gametour.com/site/editorial.php/edit/23

Davies, M. (2006, June 15). Gamers don't want any more grief. *The Guardian*. Retrieved from http://www.guardian.co.uk/technology/2006/jun/15/games.guardianweeklytechnologysection2

Dibbell, J. (2006). *Play money: Or, how I quit my day job and made millions trading virtual loot*. New York: Basic Books.

diSessa, A. A. (2000). *Changing minds: Computers, learning, and literacy*. Cambridge, MA: Massachusetts Institute of Technology Press.

Douban, G. (2007, November 27). To Nintendo's surprise, Wii is hot with seniors. *Christian Science Monitor*. Retrieved from http://www.csmonitor.com/2007/1127/p01s05-ussc.html

Dowd, R., Singer, D. G., & Wilson, R. F. (Eds.). (2006). *Handbook of children, culture, and violence*. Thousand Oaks, CA: Sage.

Duckworth, E. L., Peterson, C., Matthews, M. D., & Kelly, D. R. (2007). Grit: Perseverance and passion for long-term goals. *Journal of Personality and Social Psychology, 92*(6), 1087–1101.

Dweck, C. (2006). *Mindset: The new psychology of success*. New York: Random House.

Ericsson, K. A., Krampe, R. T., & Tesch-Romer, C. (1993). The role of deliberate practice in the acquisition of expert performance. *Psychological Review, 100*(3), 363–406.

Erlich, E. (2006). *Why America must innovate: A call to action* [Report]. Washington, DC: National Governors Association. Retrieved from http://www.nga.org/Files/pdf/0702INNOVATIONCALLTOACTION.PDF

Ehrenreich, B. (2001). *Nickel and dimed: On (not) getting by in America*. New York: Metropolitan Books.

Electronic Arts. (2009). The Sims 3: *Movies and more*. Retrieved from http://www.thesims3.com/moviesandmore

Entertainment Software Association. (2009). *Essential facts about the computer and video game industry*. Washington: Author. Retrieved from http://www.theesa.com/facts/gameplayer.asp

Ericsson, K. A., Charness, N., Feltovich, P., & Hoffman, H. R. (Eds.). (2006). *Cambridge handbook on expertise and expert performance*. Cambridge, UK: Cambridge University Press.

Fair Test. (2007, August 17). Multiple choice tests. Retrieved from http://www.fairtest.org/facts/mctfcat.html

Faylen. (2006, July 6). Faylen's skinning tutorial 1—basic clothing recolor. *Mod The Sims*. Retrieved from http://www.modthesims.info/showthread.php?t=172166

Federation of American Scientists. (2006). *Harnessing the power of video games for learning*. Washington: Author.

Fine, G. A. (1983). *Shared fantasy: Role playing games as social worlds*. Chicago: University of Chicago Press.

Friedman, T. (2005). *The world is flat: A brief history of the twenty-first century*. New York: Farrar, Straus and Giroux.

Friedman, T. L. (2008). *Hot, flat, and crowded: Why we need a green revolution—and how it can renew America*. New York: Farrar, Straus, & Giroux.

Fullerton, T. (2008). *Game design workshop, Second Edition: A playcentric approach to creating innovative games*. Burlington, MA: Morgan Kaufmann.

Gee, J. P. (1990). *Sociolinguistics and literacies: Ideologies in discourses*. London: Taylor & Francis.

Gee, J. P. (2003). *What video games have to teach us about learning and literacy*. New York: Palgrave Macmillan.

Gee, J. P. (2004). *Situated language and learning: A critique of traditional schooling*. London: Routledge.

Gee, J. P. (2007). *Good video games and good learning: Collected essays on video games, learning, and literacy*. New York: Peter Lang.

Gee, J. P. (2008). *Getting over the slump: Innovation strategies to promote children's learning*. New York: The Joan Ganz Cooney Center.

Gee, J. P., & Hayes, E. (2009). No quitting without saving after bad events: Gaming paradigms and learning in *The Sims*. *International Journal of Learning & Media, 1*(3), 1–17.

Gladwell, M. (2008). *Outliers: The story of success*. New York: Little Brown.

Glenberg, A. M. (1997). What memory is for. *Behavioral and Brain Sciences, 20*(1), 1–55.

Goleman, D. (1995). *Emotional intelligence: Why it can matter more than IQ*. New York: Bantam Books.

Goleman, D. (1997). *Social intelligence: The new science of human relationships*. New York: Bantam Books.

Goode, J., Estrella, R., and Margolis, J. (2006). Lost in translation: Gender and high school computer science. In J. M. Cohoon & W. Aspray (Eds.),

Women in information technology: Research on the reasons for under-representation (pp. 89–114). Cambridge, MA: Massachusetts Institute of Technology Press.

Gore. A. (2006). *An inconvenient truth: The planetary emergency of global warming and what we can do about it.* New York: Rodale Books.

Gorriz, C. M., & Medina, M. (2000). Engaging girls with computers through software games. *Communications of the Association for Computing Machinery, 43*, 42–49.

Gould, S. J. (1981). *The mismeasure of man.* New York: W. W. Norton.

Graff, H. J. (1979). *The literacy myth: Literacy and social structure in the nineteenth-century city.* New York: Academic Press.

Graff, H. J. (1987). *The legacies of literacy: Continuities and contradictions in western society and culture.* Bloomington, IN: Indiana University Press.

Greenblatt, S. (1983). *Renaissance self-fashioning: From More to Shakespeare.* Chicago: University of Chicago Press.

Grossman, L. (2008, April 24). Stephenie Meyer: A new J. K. Rowling? *Time.* Retrieved from http://www.time.com/time/magazine/article/0,9171,1734838,00.html

Grossman, L. (2009, June 17). Iran protests: Twitter, the medium of the movement. *Time.* Retrieved from http://www.time.com/time/world/article/0,8599,1905125,00.html

Hayes, E. (2005). Women, video gaming, & learning: Beyond stereotypes. *TechTrends, 49*(5), 23–28.

Hayes, E. (2007). Women and video gaming: Gendered identities at play. *Games and Culture, 2*(1), 23–48.

Hayes, E. (2008a). Game content creation and IT proficiency: An exploratory study. *Computers & Education, 51*(1), 97–108.

Hayes, E. (2008b). Girls, gaming, and trajectories of technological expertise. In Y. B. Kafai, C. Heeter, J. Denner, & J. Sun (Eds.), *Beyond Barbie and Mortal Kombat: New perspectives on gender, games, and computing.* (pp. 183–194). Cambridge, MA: Massachusetts Institute of Technology Press.

Hayes, E., & Games, I. (2008). Learning through game design: A review of current software and research. *Games and Culture, 3*, 309–332.

Hayes, E., & Gee, J. P. (2009). Public pedagogy through video games: Design, resources & affinity spaces. In J. A. Sandlin, B. D. Schultz, &

J. Burdick (Eds.), *Handbook of public pedagogy: Education and learning beyond schooling*. New York: Routledge.

Hayes, E. & Gee, J. P. (in press). No selling the genie lamp: A game literacy practice in *The Sims*. To appear in *E-learning*.

Hayes, E., & King, E. M. (2009). Not just a dollhouse: What *The Sims 2* can teach us about women's IT learning. *On The Horizon, 17*(1), 60–69.

Hayes, E., King, E. M., & Lammers, J. (2008). *The Sims 2 and women's IT learning*. Paper presented at the 49th Annual Adult Education Research Conference, St. Louis, MO. Retrieved from http://www.adulterc.org/Proceedings/2008/Proceedings/Hayes_King_Lammers.pdf

Hellekson, K., & Busse, K. (Eds.). (2006). *Fan fiction and fan communities in the age of the Internet*. Jefferson, NC: McFarland.

Hestenes, D. (1999). The scientific method. *American Journal of Physics, 67*(1), 274.

Hirsch, E. D. (1987). *Cultural literacy: What every American needs to know*. New York: Houghton Mifflin.

Hutchins, E. (1995). *Cognition in the wild*. Cambridge, MA: Massachusetts Institute of Technology Press.

Ito, M., Horst, H. A., Bittanti, M., Boyd, D., Herr-Stephenson, B., Lange, P. G., Pascoe, C. J., & Robinson, L. (2008). *Living and learning with new media: Summary of findings from the Digital Youth Project*. Chicago: The John D. and Catherine T. MacArthur Foundation.

Jenkins, H. (2006). *Convergence culture: Where old and new media collide*. New York: New York University Press.

Jenkins, D. (2008, April 16). *The Sims* Franchise Breaks 100 Million Sales Milestone. *Gamasutra News*. Retrieved from http://www.gamasutra.com/php-bin/news_index.php?story=18259

Jenkins, H., Purushotma, R., Weigel, M., Clinton, K., & Robison, A. J. (2009). *Confronting the challenges of participatory culture: Media education for the 21st century*. Cambridge, MA: The Massachusetts Institute of Technology Press.

Johnson, S. (2006). *Everything bad is good for you: How today's popular culture is actually making us smarter*. New York: Riverhead.

Jo Kay. (n.d.). Educational uses of Second Life. Retrieved December 23, 2009, from http://wiki.jokaydia.com/page/Edu_SL

Juul, J. (2005). *Half-real: Video games between real rules and fictional worlds*. Cambridge: The Massachusetts Institute of Technology Press.

K–12 Embodied and Mediated Learning. (2009). Retrieved from http://ame2.asu.edu/projects/emlearning/

Kasavin, G. (2004, June 4). *The Chronicles of Riddick: Escape From Butcher Bay* review for Xbox [article]. *GameSpot.* Retrieved December 20, 2009 from http://www.gamespot.com

Keefer, J. (2004, December 3) *The Lord of the Rings: The Battle for Middle-earth* [article]. *GameSpy.* Retrieved December 20, 2009, from http://pc.gamespy.com/pc/the-lord-of-the-rings-the-battle-for-middle-earth/570709p1.html

Kennedy, J. (2007, June 1). China: Liveblogging from ground zero. *Global Voices.* Retrieved from http://globalvoicesonline.org/2007/06/01/china-liveblogging-from-ground-zero/

Kiel, P. (2008, October 23). Greenspan says 'I still don't fully understand' what happened. *ProPublica.* Retrieved from http://www.propublica.org/article/greenspan-says-i-still-dont-fully-understand-what-happened-1023

Klopfer, E., Osterweil, S., & Salen, K. (2009). *Moving learning games forward.* Cambridge, MA: The Education Arcade.

Konami. (2008). Yu-Gi-Oh!'s 5D Trading Card Game. *Yugioh-card.com.* Retrieved from http://www.yugioh-card.com/en/

Kosak, D. (2005, March 14). Will Wright presents Spore . . . and a new way to think about games. *Gamespy.* Retrieved from http://www.gamespy.com/articles/595/595975p1.html

Kutner, L., & Olson, C. (2008). *Grand theft childhood: The surprising truth about violent video games and what parents can do.* New York: Simon & Schuster.

Krentz, J. A. (Ed.). (1992). *Dangerous men and adventurous women: Romance writers on the appeal of the romance.* Philadelphia: University of Pennsylvania Press.

Lakegirl. (2009). Publishing FAQ's. *Twilight Lexicon.* [Article]. Retrieved from http://www.twilightlexicon.com/?page_id=209

Lam, W. S. E., and Rosario-Ramos, E. (2009). Multilingual literacies in transnational digitally-mediated contexts: An exploratory study of immigrant teens in the U.S. *Language and Education, 23*(2), 171–190.

Lan, W. Y., & Repman, J. (1995). The effects of social learning context and modeling on persistence and dynamism in academic activities. *The Journal of Experimental Education, 64*(1), 53–67.

Lareau, A. (2003). *Unequal childhoods: Class, race, and family life.* Berkeley: University of California Press.

LaurenplayerP. (2009, March 3). love is an understatement [Message]. Retrieved from http://thesims2.ea.com/mysimpage/guestbook.php ?user_id=2877919&nstart=22

Leadbeater, C. & Miller, P. (2004). *The pro-am revolution: How enthusiasts are changing our society and economy.* London: Demos.

Lehrer, R., & Schauble, L. (2006). Cultivating model-based reasoning in science education. In R. K. Sawyer (Ed.). *The Cambridge handbook of the learning sciences* (pp. 371–387). Cambridge: Cambridge University Press.

Lee, J. (2006). *Tracking achievement gaps and assessing the impact of NCLB on the gaps: An in-depth look into national and state reading and math outcome trends.* Cambridge, MA: The Civil Rights Project at Harvard University.

Lemann, N. (2000). *The big test: The secret history of the American meritocracy.* New York: Farrar, Straus and Giroux.

Lenhardt, A., & Madden, M. (2005). *Teen content creators and consumers.* Washington, DC: Pew Internet & American Life Project. Retrieved from http://www.pewInternet.org/PPF/r/166/report_display.asp

Lemke, M., Sen, A., Pahlke, E., Partelow, L., Miller, D., Williams, T., Kastberg, D., & Jocelyn, L. (2004). *International outcomes of learning in mathematics literacy and problem solving: PISA 2003 results from the U.S. perspective* (NCES 2005-003). Washington, DC: National Center for Education Statistics.

Lenhart, A., Kahn, J., Middaugh, E., Macgill, A., Evans, C., & Vitak, J. (2008). *Teens, video games, and civics.* Washington, DC: Pew Internet and American Life Project. Retrieved from http://www.pewinternet.org/Reports/2008/Teens-Video-Games-and-Civics.aspx

Linden Lab. (2009). Virtual environments enable new models of learning. *Second Life Grid.* Retrieved from http://secondlifegrid.net/slfe/education-use-virtual-world

Linden, T. (2009, April 16). The *Second Life* economy—First Quarter 2009 in detail [Blog entry]. Retrieved from https://blogs.secondlife.com/community/features/blog/2009/04/16/the-second-life-economy--first-quarter-2009-in-detail

Lockwood, M. (2007, January 24). The Sims phenomenon. Retrieved from http://ezinearticles.com/?The-Sims-Phenomenon&id=430065

Madaus, G., Russell, M, & Higgins, J. (2009). *The paradoxes of high-stakes testing: How they affect students, their parents, teachers, principals, schools, and society.* Charlotte, NC: Information Age Publishing.

Margolis, J., & Fisher, A. (2002). *Unlocking the clubhouse: Women in computing.* Cambridge, MA: Massachusetts Institute of Technology Press.

Matti. (2008, April 19). Re: Lincoln Heights [Message 115].Retrieved from http://similik.proboards.com/index.cgi?board=stories&action=display&thread=3126&page=8

Marken, A. (2009, April 21). Hard core gamers taking a back seat. *TechLife Post.* Retrieved from http://techlifepost.com/2009/04/21/hard-core-gamers-taking-a-back-seat/

McDermott, T. (2007, July 5). High points: The best games based on movies. *Darkzero.* Retrieved from http://darkzero.co.uk/game-articles/high-points-the-best-games-based-on-movies/

Meadows, M. S. (2008). *I, Avatar: The culture and consequences of having a Second Life.* Berkeley, CA: New Riders.

Merlin. (2009, May 28). TSR has already a Workshop for CC . . . *More Awesome Than You.* Retrieved from http://www.moreawesomethanyou.com/smf/index.php/topic,15059.msg433718.html#msg433718

Metacritic. (n.d.). *Harry Potter and the Order of the Phoenix,* Xbox 360 [review summary]. Retrieved on December 20, 2009, from http://www.metacritic.com/games/platforms/xbox360/harrypotterandtheorderofthephoenix

Meyer, S. (2006, June 19). Editing: How it works. *Bella Penombra.* Retrieved from http://s15.zetaboards.com/Bella_Penombra/topic/188409/1/

Michael, D., & Chen, S. (2005). *Serious games: Games that educate, train, and inform.* Florence, KY: Course Technology PTR (Profession Technical Reference).

Microsoft. (2002). *Age of Mythology*: Home. Retrieved on December 20, 2009, from http://www.microsoft.com/games/ageofmythology/norse_home.aspx

MonkeyGalPly. (2007a, February 14). *Confessions of a teen idol/geek prologue.* Retrieved from http://thesims2.ea.com/exchange/story_detail.php?asset_id=143706&asset_type=story&user_id=2877919

MonkeyGalPly. (2007b, November 12). *Lincoln Heights 1.1: A new beginning.* Retrieved from http://thesims2.ea.com/exchange/story_detail .php?asset_id=193764&asset_type=story&user_id=2877919

MonkeyGalPly. (2008, 22 January). Hey! I <3 my mom! [Blog post]. Retrieved from http://thesims2.ea.com/mysimpage/pod.php?user_id =2877919

Mountjoy, B. (2008, April 17). The Sims game breaks 100 million. *Firefox News.* Retrieved from http://firefox.org/news/articles/1402/1/The-Sims -Game-Breaks-100-Million/Page1.html

mrskathycohen (2008, April 22). Re: The Nickel and Dimed Challenge [BBS post]. Retrieved from http://bbs.thesims2.ea.com/community/ bbs/messages.php?threadID=e3c4af5e8c79145b4290c82f3eef3429 &directoryID=128&startRow=1&openItemID=item.128,root.1,item .43,item.61,item.104,item.41#c7cd7a2abc5317831ea54923c634935e

National Center for Women and Information Technology. (2009). *Women and information technology: By the numbers.* Washington, DC: National Center for Women and Information Technology.

National Science Foundation. (2007). *Back to school: Five myths about girls and science.* Washington: Author. Retrieved from http://www.nsf.gov/ news/news_summ.jsp?cntn_id=109939

Neuman, S. B., & Celano, D. (2006). The knowledge gap: Implications of leveling the playing field for low-income and middle-income children. *Reading Research Quarterly, 41*(2), 176–201.

New Commission on the Skills of the American Workforce (2007). *Tough choices or tough times.* Washington, DC: National Center on Education and the Economy.

Ortutay, B. (2008, April 16). 'The Sims' sells 100 million copies worldwide. *MSNBC.* Retrieved from http://www.msnbc.msn.com/id/24163937/

Parfitt, O. (2008, November 12). Max Payne UK review: The classic game becomes an awful movie. *Imagine Games Network* (*IGN*). Retrieved from http://movies.ign.com/articles/929/929335p1.html

Parker, G. (2002). *Cross-functional teams: Working with allies, enemies, and other strangers.* San Francisco: Jossey-Bass.

Partnership for 21st Century Skills. (2004). *Framework for 21st century learning.* Retrieved from http://www.21stcenturyskills.org/index .php?option=com_content&task=view&id=254&Itemid=120

Partnership for 21st Century Skills. (2007). *Teaching and learning for the 21st century: Report of the Arizona summit on 21st century skills.* Retrieved from http://www.21stcenturyskills.org/documents/arizona_summit_on_21st_century_skills_report.pdf

Payne, C. (2008). *So much reform, so little change: The persistence of failure in urban schools.* Cambridge, MA: Harvard Education Press.

Pellegrino, J., Chodowsky, N., & Glaser, R. (2001). *Knowing what students know: The science and design of educational assessment.* Washington, DC: National Academy Press.

Pojo. (2009). Welcome to Pojo's Yu-Gi-Oh! site. *Pojo.com.* Retrieved from http://www.pojo.com/Yu-gi-oh/

Programme for International Student Assessment (PISA). (2006). *Science competencies for tomorrow's world: Vol. 1. Analysis.* Paris: Organisation for Economic Co-Operation and Development (OECD) Publishing.

Radway, J. A. (1991). *Reading the romance: Women, patriarchy, and popular literature.* Raleigh: University of North Carolina Press.

Reeves, B., and Nass, C. (1999). *The media equation: How people treat computers, television, and new media like real people and places.* New York: Cambridge University Press.

Reich, R. B. (1992). *The work of nations.* New York: Vintage Books.

Reich, R. B. (2001). *The future of success: Working and living in the new economy.* New York: Knopf.

Rheingold, H. (2000). *The virtual community: Homesteading on the electronic frontier* (Rev. ed.). Cambridge, MA: Massachusetts Institute of Technology Press.

Rohina. (2009, May 29). Important notice from the GRAMMAR POLICE. Plz read. This means you [Forum message]. *More Awesome Than You!* Retrieved from http://www.moreawesomethanyou.com/smf/index.php?topic=15068.0

Rymaszewski, M. (2007). *Second Life: The official guide.* Indianapolis: Wiley.

Salen, K. (Ed.). (2007). *The ecology of games: Connecting youth, games, and learning.* Cambridge, MA: The Massachusetts Institute of Technology Press.

Salen, K., and Zimmerman, E. (2003). *Rules of play: Game design fundamentals.* Cambridge, MA: The Massachusetts Institute of Technology Press.

Schiesel, S. (2006, May 7). Welcome to the new dollhouse. *The New York Times*. Retrieved from http://www.nytimes.com/2006/05/07/arts/07schi.html

Schell, J. (2008). *The art of game design: A book of lenses*. Burlington, MA: Morgan Kaufmann.

Schön, D. A. (1983). *The reflective practitioner: how professionals think in action*. New York: Basic Books.

Schwartz, D. L., Sears, D., & Chang, J. (2007). Reconsidering prior knowledge. In M. Lovett and P. Shah (Eds.), *Thinking with data* (pp. 319–344). Mahwah, NJ: Erlbaum.

Semel, S. (2002). Progressive education. In D. L. Levinson, P. W. Cookson Jr., and A. R. Sadovnik (Eds.), *Education and sociology: An encyclopedia* (pp. 463–472). New York: Routledge Falmer.

Shaffer, D. W. (2005). Epistemic games. *Innovate, 1*(6). Retrieved from http://www.innovateonline.info/index.php?view=article&id=81

Shaffer, D. W. (2007). *How computer games help children learn*. New York: Palgrave Macmillan.

Shirky, C. (2008). *Here comes everybody: The power of organizing without organizations*. New York: Penguin.

Sidekickk! (2008, April 19). Re: Lincoln Heights [Message 114]. Retrieved from http://similik.proboards.com/index.cgi?board=stories&action=display&thread=3126&page=8

Squire, K. D. (2006). From content to context: Video games as designed experiences. *Educational Researcher, 35*(8), 19–29.

Steinkuehler, C. A. (2008a). Cognition and literacy in massively multiplayer online games. In J. Coiro, M. Knobel, C. Lankshear, & D. Leu (Eds.), *Handbook of research on new literacies* (pp. 611–634). Mahwah, NJ: Erlbaum.

Steinkuehler, C. (2008b). Massively multiplayer online games as an educational technology: An outline for research. *Educational Technology, 48*(1), 10–21.

STEM Education Coalition (2009). Coalition objectives. Retrieved from http://www.stemedcoalition.org//content/objectives/

Stevens, A. (2006, October 11). Evoking lives struggling to exist on bare minimums. *The New York Times*. Retrieved from http://theater2.nytimes.com/2006/10/11/theater/reviews/11nick.html

Taylor, T. L. (2006). *Play between worlds: Exploring online game culture.* Cambridge, MA: The Massachusetts Institute of Technology Press.

Terdiman, D. (2009). *Sims 3* sets franchise sales record. *CNET News: Gaming & Culture.* Retrieved from http://news.cnet.com/8301-10797_3-10261044-235.html

The Lost Legion. (2004, December 24). [Description of archived file]. Retrieved from http://aom.heavengames.com/scendesign/spotlight

Tillberg, H. K., & Cohoon, J. M. (2005). Attracting women to the CS major. *Frontiers: A Journal of Women Studies, 26*(1), 126–140.

Toffler, A., & Toffler, H. (2006). *Revolutionary wealth: How it will be created and how it will change our lives.* New York: Knopf.

Tough, P. (2008). *Whatever it takes: Geoffrey Canada's quest to change Harlem and America.* New York: Houghton Mifflin Harcourt.

Wallis, C., & Steptoe, S. (2006). How to bring our schools out of the 20th century. *Time Magazine, 168*(25), 50–56. Retrieved from http://www.time.com/time/magazine/article/0,9171,1568480,00.html

Wardrip-Fruin, N., & Harrigan, P. (Eds.). (2004). *First person: New media as story, performance, and game.* Cambridge, MA: The Massachusetts Institute of Technology Press.

Wasik, B. (2009). *And then there's this: How stories live and die in viral culture.* New York: Viking.

Wenger, E. (1998). *Communities of practice: Learning, meaning, and identity.* Cambridge: Cambridge University Press.

Wenger, E., McDermott, R., & Snyder, W. M. (2002). *Cultivating communities of practice.* Cambridge, MA: Harvard Business School Press.

Willingham, D. T. (2009). *Why don't students like school? A cognitive scientist answers questions about how the mind works and what it means for the classroom.* New York: Jossey-Bass.

Wittgenstein, L. (2001). *Philosophical investigations* (3rd Edition). Oxford: Blackwell Publishing.

Wolf, M. J. P. (Ed.). (2002). *The medium of the video game.* Austin, TX: University of Texas Press.

Wright, W. (2006). Dream machines. *Wired Magazine, 14*(04). Retrieved from http://www.wired.com/wired/archive/14.04/wright.html

Yamx. (2008, April 2). The *Nickel and Dimed* challenge [BBS post]. Retrieved from http://bbs.thesims2.ea.com/community/bbs/messages.php?threadID=e3c4af5e8c79145b4290c82f3eef3429&directoryID=128&startRow=1&openItemID=item.128,root.1,item.43,item.61,item.104,item.41

Yu-Gi-Oh Wikia. (n.d.). [Web site]. Retrieved August 4, 2009, from http://yugioh.wikia.com/wiki/Main_Page

Zane, R. (2005, August 9). Ziff Davis video game survey: Gamers continue to cut TV viewing [Press release]. Retrieved from http://www.ziffdavis.com/press/releases/050809.0.html

Index